# SpringerBriefs in Business

More information about this series at http://www.springer.com/series/8860

Masayuki Matsui

# Fundamentals and Principles of Artifacts Science

## 3M&I-Body System

 Springer

Masayuki Matsui
The University of Electro-Communications
Chofu, Tokyo
Japan

and

Faculty of Engineering, Department
   of Industrial Engineering
   and Management
Kanagawa University
Yokohama, Kanagawa
Japan

ISSN 2191-5482          ISSN 2191-5490   (electronic)
SpringerBriefs in Business
ISBN 978-981-10-0472-8          ISBN 978-981-10-0473-5   (eBook)
DOI 10.1007/978-981-10-0473-5

Library of Congress Control Number: 2016930055

Printed on acid-free paper

This Springer imprint is published by SpringerNature
The registered company is Springer Science+Business Media Singapore Pte Ltd.

# Preface

The first ever trial on a simple body was performed by Archimedes. For complex bodies, it has been over hundred years, since any study was conducted on artifacts such as the 3M&I-body system in IE/OR (industrial engineering/operations research). Generally, "3M" means to human, material/machine, money, and "I" means the information/method in nature versus artifacts-based views.

The complex subject of 3M&I variety was first introduced and discussed in the book, "Manufacturing and service enterprise with risks," by Springer (2008, 2014), and IE/OR is regarded and developed as an art (knowledge and "waza"/skill) of the 3M&I-body system in the book. F.W. Taylor introduced and developed the motion and time method and labor-related science in IE/OR, and the method (information) was further developed by N. Wiener.

For economic artifacts, "3M" represents human (labor), material/machine (manufacturing process), money (capital), and "I" represents information (wisdoms) on methods (combinatorics). This combination alone is not sufficient to delivers value (weight/profit), and it is born out of the changing state of activity and flow in the field (market/society), in contrast with nature.

This is the first book covering original knowledge on the mathematical science of the 3M&I-body system, and presents axioms and two major propositions on issues surrounding nature versus artifacts, and not on animal versus machine. This book is the product of industrial engineering as contrasted with Wiener's cybernetics challenges for over a half-century.

For the 3M&I-body, there are two approaches to systemization and control: AI (artificial intelligence)/IoT (Internet of Things) and Matsui's matrix/3D. The former is the analogical and visual approach to a real entity. The latter is the digital and logical approach to system decision, and is applied to the robotics of bodies.

The mathematical science of a body is constructed more effectively using algebra, geometry, analysis, and control, as in Matsui's equation, toward the sandwich and balancing principles of bodies. The sandwich issues propose the squeeze or pinching theorem in mathematics for the 3M&I-body system, and the balancing issues propose the principle of balancing and invisible collaboration of bodies, beginning from the work of Archimedes.

The book can contribute to the integration of knowledge and intelligence in nature versus artifacts science, and facilitate the realization of the cyber/real world, such as an enterprise robot, cloud-coordinated SCM (supply chain management) and smart cites in the near future. In addition, the book, together with the Appendix, will be useful to graduate students and researchers dealing with problems related to objects that occur in nature (Newton's) and to artifacts (Matsui's).

In conclusion, the author would like to acknowledge the helpful and valuable advice given by Dr. Kiyomasa Narita, Professor Emeritus of Kanagawa University, and thank Dr. Midori Mori, Associate Professor, and Yusuke Sakai, a member on the technical staff, for editing this book. Further, the author wishes to thank the faculty of Kanagawa University, for their support toward this study.

Finally, the author appreciates his dear wife, Kazuko, and family members for their help and support in his academic life.

Tokyo, Japan                                                          Masayuki Matsui
2015

# Contents

# Chapter 1
# Introductory 3M&I-Body Problem

**Abstract** Industrial Engineering and Operations Research (IEOR) could be regarded as the art of 3M&I-body systems, in which 3M is the human, material/machine, and money and I is the information. This chapter reviews the IEOR versus Wiener's cybernetics issues, and relates sandwich and balancing propositions to the nature versus artifacts body of research. The former (sandwich) would probably be caused by the dualism (double property) of Matsui's, Newton's, and Jewell's equations, and the latter (balancing) originates in the invisible body-balancing principle and economics shown by the medium criteria in supply chain management (SCM)/Smart city. There are also two approaches of artificial intelligence (AI)/Internet of Things (IoT) and Matsui's matrix/3D, but only the latter will be focused on and developed in this book. Finally, a review of the tendency toward post-ERP/SCM is given for higher management. For example, the enterprise resource planning (ERP) may be partially replaced by CEO or enterprise robots, and SCM will probably be replaced by cloud balancing in the near future.

**Keywords** 3M&I · Cybernetics · Nature versus artifacts · Sandwich · Balancing · IEOR

## 1.1 Industrial Engineering and Operations Research Versus Wiener's Cybernetics View

### 1.1.1 Introduction

The subject of 3M&I variety has been recently introduced and systematically discussed by Matsui [1]. Generally, IEOR could be regarded as the art of 3M&I-body systems, while Wiener's cybernetics [2] would be the discussion of animal and machine in 1948. These analogies and similarities have motivated the academic study of the art of nature (animal) and artifacts (machine) in the 3M&I-body system ever since Simon's work on artificial facts [3].

The scientific study of body (object) would probably go back to Archimedes' (B.C. 287–212) studies of the balancing and area method on geometric configurations

© The Author(s) 2016                                                                                      1
M. Matsui, *Fundamentals and Principles of Artifacts Science*,
SpringerBriefs in Business, DOI 10.1007/978-981-10-0473-5_1

such as the lever, specific gravity, and squeezing. His seminal concepts of body developed toward the sandwich and balancing issues and the post-Taylor's science of labor on 3M&I-body in this book.

This first chapter presents the foundation of the 3M&I-body system, and gives an outline of the nature (Newton's) versus artifacts (Matsui's) problem, together with Appendix A.1. Next, the two main propositions of the sandwich and balancing theories are discussed, and are given new applicability to the nature versus artifacts science.

Finally, the two approaches of AI/IoT and Matsui's matrix/3D are introduced, and are reviewed with consideration of the post-ERP/SCM artifacts in the near future.

### *1.1.2   Outline of the 3M&I-Body System*

For enterprise entities, there is the 3M&I system, in which 3M is the resource of human, material/machine, and money, and I refers to information resources, as shown in Fig. 1.1. In the diagram, while not displaying the concept of value (weight/profit), this value originates from state (flow/activity) changing in the field (market/society).

This system begins with Archimedes' work (see Appendix A.2), and is seen in the studies of physics, electronics, economics, management, and so on. This is summarized in Table 1.1, and is formulated by Matsui's matrix equation [4]; it can

**Fig. 1.1** 3M&I studies in enterprise (body)

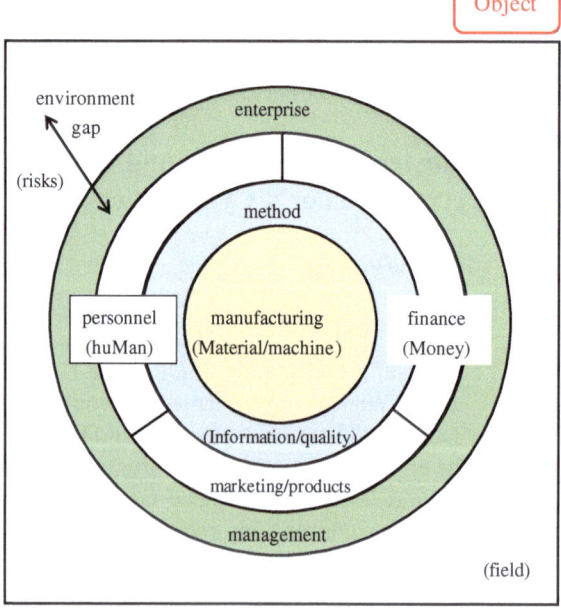

**Table 1.1**  Summary of the 3M&I-body system

| Body (3M&I) | | Economics | Particle physics | Electric motor | Characteristic equation |
|---|---|---|---|---|---|
| Resource | Human material/ Machine money | Labor manufacturing process capital | I–III generation | Resistance motor electric current | Operator |
| Method | Information | Business | Particles that convey the force | Control | |
| Goal | Profit | Marginal profit | Higgs boson (weight) | Position | Proper state (proper vector) |
| | Lead time | Interests | Graviton (gravity) | Speed | |

be turned into a characteristic equation type of problem. This originates in sandwich theory, and could be generalized to a proposition in the structure of 3M&I-body systems.

### 1.1.3   Cybernetics Versus 3M&I Issues

Wiener's cybernetics concept is related to the animal and machine in its methods of communication and control, while 3M&I science is related to the natural and artifacts body in the multi(N)-dimensional space (Fig. 1.2 [5]).

In Fig. 1.2, the orthogonal projection to the vector $y(\in V_N)$ is available as follows:

$$y_F^{\perp} = y - Fa, \quad y_F = Fa, \tag{1.1}$$

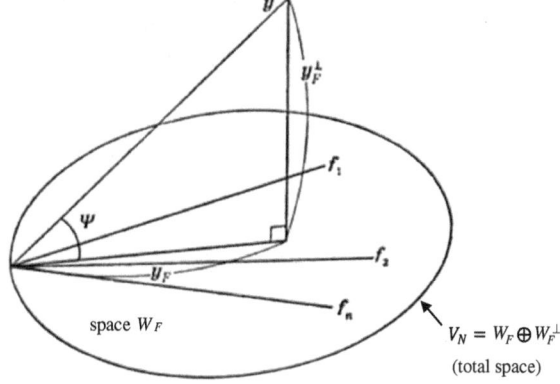

**Fig. 1.2**  Multidimensional image of the 3M&I-body system

**Table 1.2** Cybernetics versus 3M&I science

| Type | Wiener's type | 3M&I type |
|------|---------------|-----------|
| Object | Animal versus machine | Natural versus artifacts |
| Structure | Communication and control | Matsui's equation (dualism) |
|  | Feedback | Sandwich |
| Value | *Muda*[a] versus value | Efficiency versus *muda*[a] |
|  | Entropy | Chameleon's criteria |
| Input–output | Arithemetic mean | (stochastic) Medium |
|  | Pair(series) body | Pair(OE) body |
| Series–parallel | Circuit/neuropil | Conveyor/SCM |
|  | Dual system | Balancing |

[a]"muda" means the redundancy, useless, waste, delay, etc.

where $F = (f_1, f_2, \ldots, f_p)$ and $a$ is weight (coordination) vector at $(N, p)$-position in multivariate management.

The knowledge forms of the 3M&I system are Matsui's equation and the chameleon's criteria, in which "chameleon" means the color (redundancy)-changing animal (body), in Matsui [4]. Those of Wiener's cybernetics are feedback and entropy. The correspondence of both sciences can be summarized as shown in Table 1.2, in structure, value, input–output, and series/parallel types. From these discussions, the two main propositions of sandwich and balancing theory are found in 3M&I science.

### 1.1.4  Value Concept in the 3M&I-Body System

The so-called value is a combination of usefulness and scarcity, but the former is not seen in cybernetics. That is, the scarcity is based on the amount of information (entropy), and usefulness is not considered (Fig. 1.3).

Table 1.3 shows the respective values in cycle time, revenue, and worth. Usefulness $(L)$ is related to the realization (efficiency) of resources events, the scarcity $(Z)$ is related to the demand and supply (delay/*muda*) of events in information, and the resulting value could be calculated with Matsui's $W$ $(=ZL)$ as follows:

$$(\text{Value}, W) = (\text{Scarcity}, Z) \times (\text{Usefulness}, L). \tag{1.2}$$

**Fig. 1.3** Sandwich by the progressive figures in electronics (Jewell's vs. Matsui's in cumulative diagram)

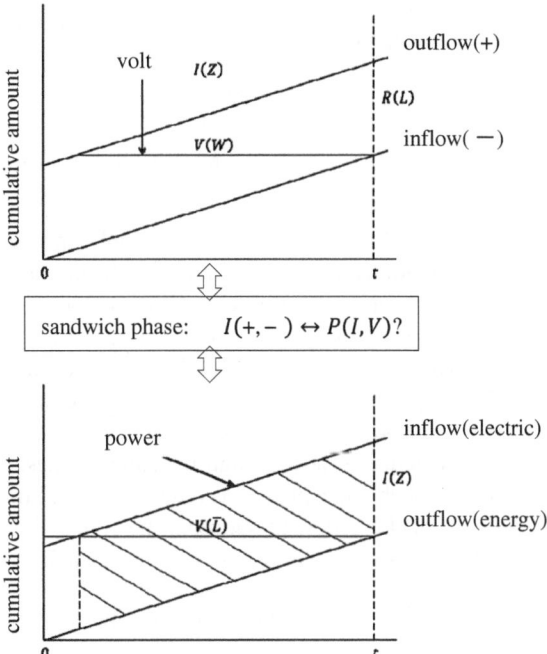

**Table 1.3** What is value?: (Usefulness) × (Scarcity)

| Value | Usefulness (L) | | Scarcity (Z) | |
|---|---|---|---|---|
| Cycle time (Z) $\lambda Z = 1+\eta$ | Delay time | D | Processing time (overflows, η) | X |
| | Busy time | | Repair time | |
| Revenue (ER) ER = EN + EC | Labor theory of value | EN (marginal profit) | Subjective theory of value | EC (variable cost) |
| | Surplus value | | Marginal utility | |
| Worth (W) W = ZL | Chameleon's criteria | $\bar{\beta}$ (medium) | Entropy | H |
| | Capacity utilization | $u(\rho)$ | Redundancy | R |

## 1.2    New Challenge: Nature Versus Artifacts Issues

### 1.2.1    Creative Challenge and Finite Mathematics

There is a new challenge creating the science of 3M&I-body systems. Creativity is required in "systemization" which is the design involved in assembling the separate objects, or making connections between two intelligences as shown by J. H. Poincare' (1854–1912).

**Table 1.4** Dualism of main formulas: Newton, Jewell, and Matsui

| Dualism | Lower level | Upper level |
|---------|-------------|-------------|
| Matsui's | $W = ZL\ (=\overline{L})$ | $\overline{W}=\overline{ZL}\ (=Z^2L)$ |
| Newton's | $p = mv$ | $F = ma\ (=mv^2)$ |
| Jewell's | $V = IR$ | $P = IV\ (=I^2R)$ |

(For Matsui's vs. Newton's case, see Appendix A.1)

Our creative finding is the dualism (double property) in Newton's, Jewell's, and Matsui's equations, shown in Table 1.4. In the table, the lower level refers to the scalar, and the upper level refers to the vector (phase). (Later, Fig. 1.4 shows the dualism corresponding to the sandwich model and 3M&I-body systems.)

The creativity required for this is formalized in a finite mathematical series by J. G. Kemeny and J. L. Snell [6]. Finite mathematics has become a key tool in the art of 3M&I-body systems since the twentieth century, and is also useful to our creative challenge: sandwich and balancing issues.

The sandwich concept, in this case, would be similar to the squeeze or pinching theorem in mathematics. In addition, the concept of balance originates in the lever principle in physics, and its meaning is similar in form to Matsui's equation as follows:

$$\text{Moment of power } (W) = Z_1 L_1 = Z_2 L_2. \tag{1.3}$$

### 1.2.2  Outline of Sandwich Theory (Proposition)

This sandwich (S = W) theory originated in the 2-center model in 1983, and the S = W theory of enterprise was first sketched in 2009, as shown in Fig. 1.4. Figure 1.4 shows the (evolutional) sublation to the dilemma of bottom-up versus top-down approaches to management in the organization.

**Fig. 1.4** Sandwich theory (S = W Theory) of enterprise: A set view (*"waist" means the middle or central part in (human) body **"Jyukyu": demand($d$)-to-supply($m$))

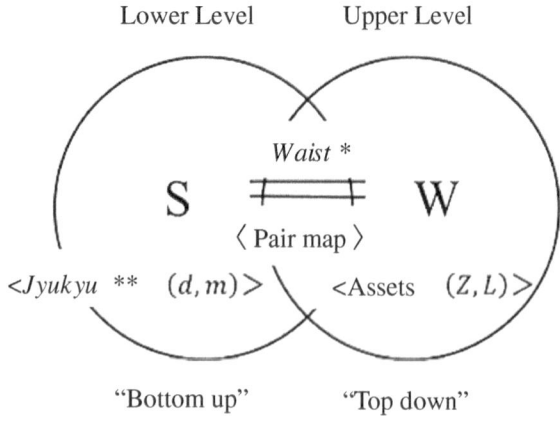

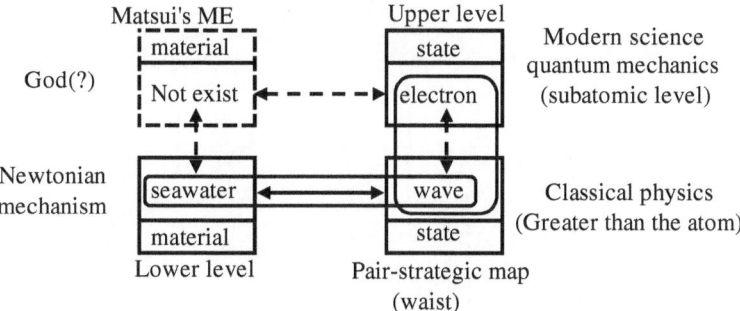

**Fig. 1.5** Sandwich theory: a physical view and artifacts scheme

That is, this has led to the shift change, from the bottom-up and top-down toward middle–central approach, to operations/decisions in enterprise. The result is, for example, seen in the use of Nash's solution to the pair map [7] as follows:

$$(Z+L)/2 \geq \{Z = L\} \geq \sqrt{ZL}. \tag{1.4}$$

For physics, an application of the S = W theory can be seen in Fig. 1.5. The seawater (material) and electron (state) exist, respectively, in the lower and upper levels of sandwich, while the wave (state) exists in the middle area (pair map). At the top of the body, god-like Matsui ME might reign over the upper level or the sandwich itself.

This type of the problem could, for example, be formulated by the independent component analysis (ICA) type of Matsui's matrix equation (Matsui's ME) [4]. The formulation is the vector-control motor (artifacts) [8] discussed in a later chapter. These problems generally depend on and would be solved with the characteristic equation in nature versus artifacts.

### 1.2.3   Outline of Balancing Theory (Proposition)

The balancing theory originates in the problem of assembly line or conveyor system type, and is critical to the integral balancing or invisible collaboration [1]. Recently, this is geared toward the supply chain [4, 9] and smart city [10], under the global society/economics [11]. Illustrative figures are seen in Figs. 1.6 and 1.7.

This balancing principle and stability is hypothesized independently using the principle of the lever (1.3) as follows [11]:

$$a_1\bar{\beta}_1 = a_2\bar{\beta}_2 = \cdots = a_n\bar{\beta}_n = W(=ZL), \tag{1.5}$$

in which $\bar{\beta}'_i s$ are Chameleon's criteria, $a'_i s$ are weighting factors, and Matsui's $W(=ZL)$ is the balancing value (stability) at its equilibrium.

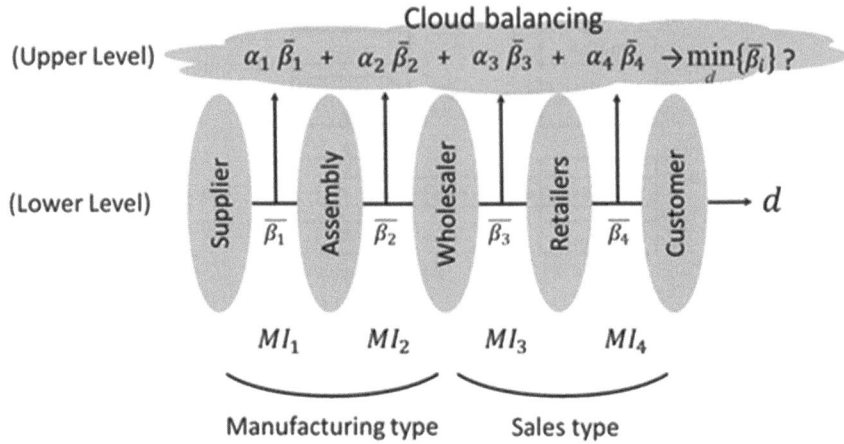

**Fig. 1.6** Body-balancing system of a supply chain (economics) [4, 9]

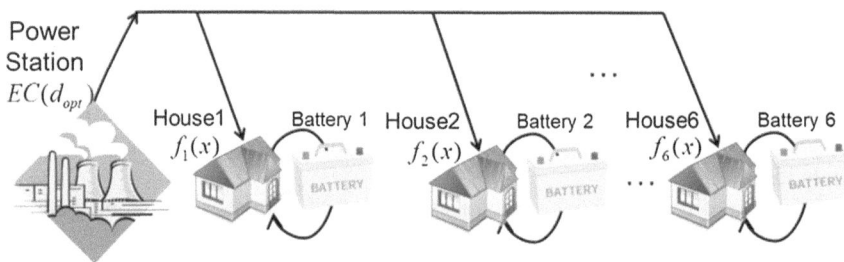

**Fig. 1.7** Basic model of a smart city [10]

This principle relates its factors in a series type arrangement (Fig. 1.6); a corresponding parallel type (Fig. 1.7) has not been studied. Probably, it would be related to a min–max formulation.

As another medium approach, the following balancing (win–win) principle is hypothesized independently as follows:

$$EN_1 = EN_2 = \cdots = EN_n, \quad EN = ER - EC \tag{1.6}$$

for series type, while for parallel type,

$$b_1 EC_1 = b_2 EC_2 = \ldots = b_n EC_n = W(=ZL), \quad EC = ER - EN. \tag{1.7}$$

where $b_i's$ are weighting factors.

It is noted that Eqs. (1.5) and (1.6) are related to a dual condition in reliability theory, resulting in the maximum total profit (strength).

**Table 1.5**  Comparative methods and tools for higher management

| Method | | AI/IoT | Matrix/3D |
|---|---|---|---|
| Approach | | Analog type | Digital type |
| Visualization | | Monitor type | Logic type |
| Real time | | Δ | Δ |
| System decision | | Δ | ○ |
| Hyper speed | | ○ | ○ |
| Tool | ERP | Socrates, CEO robot (naming) | Enterprise-motor body (logic) |
| | SCM | *momo-mo* plan (idea) | Cloud SCM (logic) |

## 1.2.4   Toward Post-ERP/SCM

These are two approaches to higher management: artificial AI/IoT (analogy) [12, 13] and Matsui's matrix/3D (digital) [5, 14]. The comparisons are seen in Table 1.5, and would be fully realized by real-time robots in management.

Recently, Socrates, the CEO robot (naming by Wollmann et al. [13] in 1915) and an enterprise-motor body [15] have been introduced for enterprise-robot or post-ERP. In addition, for post-SCM, the corresponding *momo-mo* plan for e-economics (by T. Suguro 2015) and Cloud-coordinated SCM/GDP [4, 9–11] have been proposed. Using these would further develop the theory of nature versus artifacts type, and usher in the new cybernetics.

## References

1. Matsui, M. (2008). *Manufacturing and service enterprise with risks: A stochastic management approach.* Springer.
2. Wiener, N. (1948). *Cybernetics: Control and communication in the animal and the machine* (2nd ed.). NewYork: Wiley.
3. Simon, H. A. (1969). *The sciences of the artificial.* Massachusetts: MIT Press.
4. Matsui, M. (2014). *Manufacturing and service enterprise with risks II: The physics and economics of management.* Springer.
5. Takeuchi, K., & Yanai, H. (1972). *Foundations of multivariate analysis -A method by projection to linear space.* Tokyo: Toyo Keizai Shinpousha. (in Japanese).
6. Kemeny, J. G., Schleifer, A., Jr. Snell, J. L., Thompson, G. L. (1962). *Finite mathematics with business applications.* Prentice-Hall.
7. Matsui, M. (2013). An enterprise-aided theory and logic for real-time management. *International Journal of Production Research, 51*(23, 24), 7308–7312.
8. Krishnan, R. (2010). *Permanent magnet synchronous and brushless dc motor drives.* CRC Press.
9. Matsui, M. (2010). Division of work, stochastic (re-Balancing) and demand speed: From assembly line toward demand chain. *Journal of Japan Industrial Management Association, 60* (6E), 324–330.
10. Takanokura, M., Matsui, M., Tang, H. (2014). Energy management with battery system for smart city. In *Proceedings of 33rd Chinese Control Conference, Nanjing, China.*

11. Matsui, M. (2015). The invisible body-balancing economics: A medium approach. *Theoretical Economics Letters, 5*, 65–73.
12. Simon, H.A. (1977). *The new science of management decision.* Prentice-Hall.
13. Wollmann, D., & Steiner, M. T. A. (2014). Complex adaptive systems integrating the decision making process in industrial companies: A scientific conceptual model. *Applied Mechanics and Materials, 670–671*, 1601–1607.
14. Matsui, M., Ishii, N., Ohba, M., Yamada, T. (2013). About matrix approach to enterprise problems toward higher management. Preprints of Japan Industrial Management Association, Spring, 2–4 (in Japanese).
15. Matsui, M. (2015). *Sandwich theory of 3M&I-Body (1)&(2).* Spring and Fall: Preprints of Industrial Management Association. (in Japanese).

# Chapter 2
# Fundamentals of 3M&I-Body System

**Abstract** This chapter describes the mathematical foundations of the 3M&I-body system. The concept of pair matrix was first proposed in 1983 and has been explored and developed into the IO-matrix system with input (revenue) and output (cost/profit). This generalization formed the axiom system of body in vector space, and was applied to a class of Matsui's equation and Matsui's matrix equation (Matsui's ME). Based on this pair matrix, the Sandwich model (S = W model) was proposed in 2009, and was formulated as a type of Matsui's ME for the cases of enterprise, motor and particle physics bodies. The Matsui's ME of strategic plan (IDTCBG) type could be generally transformed into the corresponding characteristic equation, and was solved analytically by using the matrix computing method. Finally, the problem of sandwich control was considered by using the progressive method, and 3D dynamism was explored by using the vector cell approach for solving the Matsui equation. The progressive method was extended to the state equation of progressive control with Matsui's logic of second-order (acceleration) type. Also, the 3D dynamism was obtained by mapping the $p$-dimension to the $p/2$ ($Z$ and $L$ domain) control problem in the multidimensional space.

**Keywords** Pair matrix · Input/output · Matsui's equation · Sandwich · Characteristic equation · 3D management

## 2.1 Algebra: Pair Matrix Body and Line Versus OE

### 2.1.1 Introduction

The modeling problem of bodies is probably the inverse approach to the reductionism proposed by Rene Descartes (1596–1650), and this has been called, not the system-centered, but the body-centered approach since 1977 [1]. Here, we discuss the pair relations (move/flow) of the input-output system of bodies.

For each body, there is one input (inflow) and two outputs (outflow and overflow) in the move/flow system (line). The pair relations are of two types: (inflow,

© The Author(s) 2016
M. Matsui, *Fundamentals and Principles of Artifacts Science*,
SpringerBriefs in Business, DOI 10.1007/978-981-10-0473-5_2

outflow) and (inflow, overflow), and these are denoted as the pair matrix of line versus OE (ordered entry) type.

The latter (the OE type) is new and was first treated as the pair matrix, whereas the former (the line type) is ordinal and was later compared with this type as the opposite of pair matrices. This notation was introduced in 1983 [1, 2], and was developed into a novel approach for describing the input-output system, which was different from the production and control system in IEOR.

In Sect. 2.2, the sandwich proposition [3] of a body is discussed by using pair (matrix) approach, and is applied to the enterprise, motor and physics (body) scenarios. For this proposition, the formulations by the pair map geometrics, vector space, or Matsui's ME [4, 5] were implemented and found to be useful.

In Sect. 2.3, the problem of sandwich control is considered by using the progressive method, and the issue of 3D dynamism is explored by using the vector cell approach to Matsui's equation [2, 6]. Matsui's logic of second-order (acceleration) type and the mapping control to $p/2$ (Z and L domain) from $p$-dimension space is also discussed.

## 2.1.2  Pair Matrix: Definition and Axiom

Let us denote the set of pair matrices by $K$, and their elements by $A(a, c)$ and $B(b, d) \in K$, in which $a, b$ are inputs and $c, d$ are outputs. Then, the mathematical construct of pair matrices is introduced as follows:

(1)  Additivity axiom

For any $(A, B) \in K$, the operation of addition is given as follows:

$$A + B = ((a + b, c + d)). \tag{2.1}$$

wherein the associative and commutative laws hold. Also, for any $(A, B) \in K$,

$$^{\exists}X \in K \quad \text{s.t.} \quad A + X = B, \tag{2.2}$$

and the zero element is [0, 0].

(2)  Multiplicative axiom

Several multiplicative axioms are formulated. For $x = (a, c) \in A$ and $y = (b, d) \in B$, one axiom considers the conditional reward as posterior probability. Now, the following relations hold:

$$r(x \cap y) = r(x)r(y|x) \tag{2.3}$$

$$r(y \cap z) = r(z)r(y|z) \tag{2.4}$$

Then, the respective rewards are as follows:

$$r(y|x) = y/x, \quad r(z|y) = z/y \tag{2.5}$$

These additive and multiplicative laws hold respectively as follows:

$$r((y + z|x)) = r(y|x) + r(z|x), \tag{2.6}$$

$$r(z|x) = r(y|x)r(y|x). \tag{2.7}$$

Another axiom considers the type of pair comparison, as seen in the case of Matsui's matrix equation case [4, 5]. This type transforms the pair matrix into the weighting factor by using the analytic hierarchy process (AHP) [7].

### 2.1.3 Vector Case and Matsui's ME

Next, let us assume the elements, $x$ and $y \in K$, as pair vectors. Then, the inner and outer products are defined as follows:

$$(A, B) = ([x, y]), \quad \text{inner product} \tag{2.8}$$

$$[A \times B] = (|x \times y|), \quad \text{outer product.} \tag{2.9}$$

For Matsui's equation ($W = ZL$), this notation could give the respective generalization to $W' = Z'L'$ with angle, $\theta$. Then, this angle could be used to strategically manipulate the sandwich body.

That is, the vector and scalar products in Fig. 2.1 are as follows:

$$\overline{W}' = |x \times y| = |Z' \times L'| = |Z'||L'| \sin\theta. \tag{2.10}$$

$$W' = (x, y) = (Z', L') = |Z'||L'| \cos\theta. \tag{2.11}$$

Especially for the vector product, when $\theta = \pi/2$, $\overline{W}'$ coincides with $W$ in Matsui's equation. Also, when $\theta = 0$ ($Z = 1/\lambda$), the latter becomes Little's formula: $\lambda W = L$.

For example, let us denote the vectors as: $Z'$ and $L'$, by the revenue: as $ER$, and lead time: as $LT$, respectively, in Fig. 2.1. Then, the vector product, $W'$, denote the assets (force in physics). Also, let us denote the profit, $EN$, by the revenue, $ER$, and cost, $EC$. This scalar product is the alternative (multiplicative) definition of the profit equation, $EN = ER - EC$.

**Fig. 2.1** Matsui's equation
(W = ZL) and its
generalization (W' = Z'L')
with $\theta = \angle Z'L'$

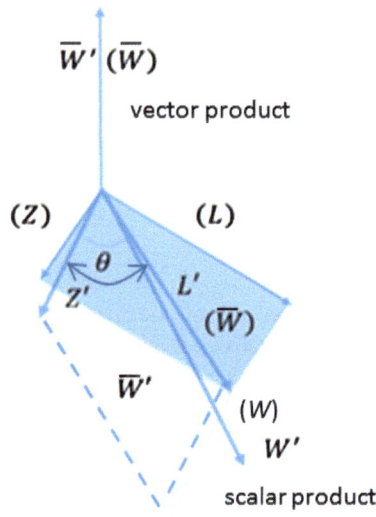

## 2.1.4  Pair Body: Line Versus OE

Here, the set of pair matrices is called the pair body. There are two types of series
bodies in the ordinal line (a) and OE bodies in the ordered-entry line (b) in Fig. 2.2.
The former has the arrival in input and departure in output, while the latter has the
arrival in input and overflow in output.

**Fig. 2.2** Pair body: line
versus OE (ordered entry)

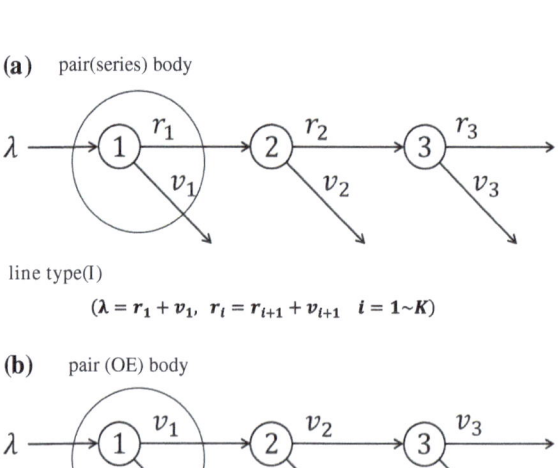

**Table 2.1** Two pair bodies and their duality

| Pair body | Series | | OE (ordered entry) | |
|---|---|---|---|---|
| Input-output | Arrival-departure | | Arrival-overflow (delay) | |
| Overflow (ratio) | Overflow (delay) | Overflow rate | Departure | Departure rate (production rate) |
| Evaluation | Addition | Multiplicative | Substraction (inner product) | Division |
| Combination | Modular | | Integral | |

Both represent the interchange relation of production ($r$) and overflow ($v$) rates, and are called the dual system. Table 2.1 lists the dual relation of line versus OE.

## 2.1.5  Duality of Series Versus OE

From Fig. 2.3, the case where the overflow rate is $v_i = 0$ consists of the transformed bodies in the modular class (additive and multiplicative). For the case in which the overflow rate $v_i \neq 0$ consists of the transformed bodies in the integral class (subtraction and division) in Fig. 2.4.

In Fig. 2.3, the revenue, $ER$, in the input, and operating (variable) cost, $EC$, in the output, are shown. In Fig. 2.4, the revenue, $ER$, in the input, and the profit, $EN$, in the output, are shown. Then, the probability of processing, $P(0 < P < 1)$, is here regarded as the $EN$ to $ER$ ratio, and the probability of loss, $B(0 < B < 1)$, is regarded as the $EC$ to $ER$ ratio.

The probability $B$ is given by

$$B = (ER, EC) = R(ER \cap EC|ER) = EC/ER, \tag{2.12}$$

where $ER \cap EC = EC(\subset ER)$. The probability P is given by

$$P = (ER, EN) = R(ER \cap EN|ER) = EN/ER, \tag{2.13}$$

where $ER \cap EN = EN(\subset ER)$.

**Fig. 2.3** Each unit of pair (series) body

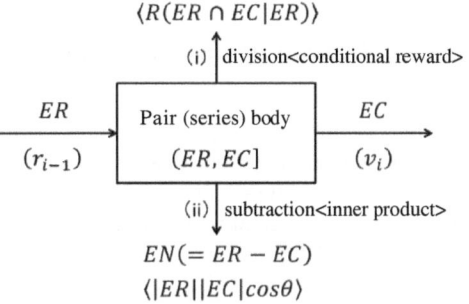

$\langle R(ER \cap EC|ER) \rangle$

(i) division<conditional reward>

$ER$     Pair (series) body     $EC$

$(r_{i-1})$     $(ER, EC]$     $(v_i)$

(ii) subtraction<inner product>

$EN (= ER - EC)$

$\langle |ER||EC|cos\theta \rangle$

**Fig. 2.4** Each unit of pair
(OE) body

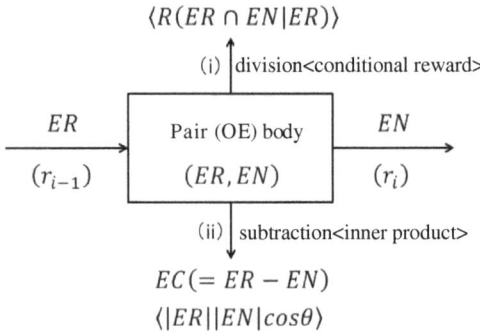

Both classes can be unified in a modular class after proper subtraction and division. For the modular class, the production ($r_\mathrm{I}$) and overflow ($r_\mathrm{II}$) rates are given as follows [8]:

$$r_\mathrm{I} = r_k = \lambda \prod_{i=1}^{k} P_i, \quad \text{line type} \tag{2.14}$$

$$v_\mathrm{II} = v_k = \lambda \prod_{i=1}^{k} P_i, \quad \text{OE type} \tag{2.15}$$

Also, the flow times, $F_\mathrm{I}$ and $F_\mathrm{II}$, correspond to the total cost and profit, respectively, and are given as follows [8]:

$$F_\mathrm{I} = \sum_{i=1}^{k} EC_i, \quad \text{cost } (\textit{linetype}) \tag{2.16}$$

$$F_\mathrm{II} = \left( \sum_{i=1}^{k} EN_i \right) \Big/ r_\mathrm{II}, \quad \text{profit } (\textit{OEtype}) \tag{2.17}$$

## 2.2   Analytics: Sandwich Theory and Matsui's Equation

### 2.2.1   Outline of the Sandwich Problem

Using pair matrix notation, which we have already developed in 1983, the management game model (MGM), consists of two centers (sales and manufacturing) [1, 9]. The MGM pair matrix table has yielded several (pair) strategic maps to the main three types of manufacturing and service enterprises, and is summarized in Fig. 2.5 [1, 3].

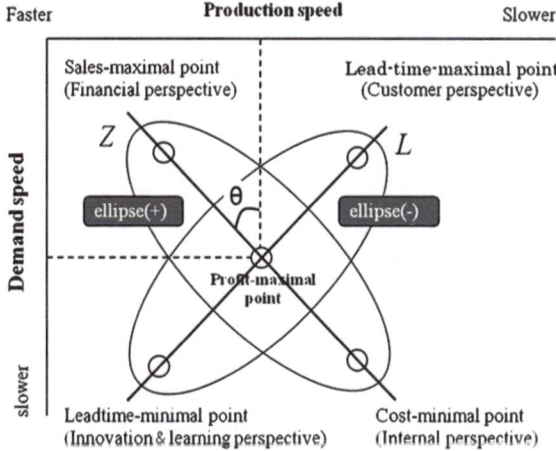

**Fig. 2.5** The balancing solution for the sandwich problem (pair map)

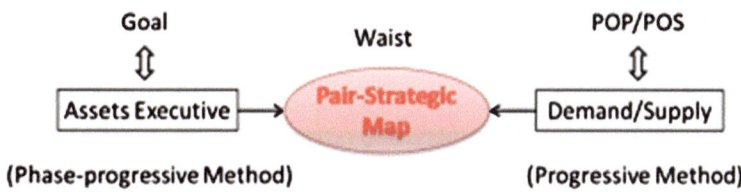

**Fig. 2.6** The sandwich model for an enterprise (toward nature vs. artifacts)

Moreover, for these pair maps, we have proposed the sandwich concept of enterprises in 2009, and have presented the sandwich model of enterprises in Fig. 2.6 (2013) [3]. This sandwich model will be developed here for the sandwich proposition of nature versus artifacts body.

For the class of 3M&I-body, our sandwich proposition is defined by the following considerations:

**Definition 2.1** The input-output system that holds the sandwich body is defined by the M-system, and the pair bodies on M-system are denoted by M-pair body.

**Proposition 2.1** *It seems sufficiently plausible that*

  (i)   *if some input-output system is assumed under a 3M&I-body, there is at least one pair body on it,*
 (ii)   *any sandwich body could be composed of an M-pair body or decomposed into an M-pair body and other, and*
(iii)   *the sandwich body could be formulated as the type of Matsui's ME.*

The meaning and validity of this proposition will be discussed for the three types of enterprise, motor and physics from the three formulations of pair map geometrics, vector space, and Matsui's ME. Then, the duality of Matsui's, Newton's, and Jewell's equation in Table 1.4 will be formalized and visualized from the mathematical aspect.

## 2.2.2  Sandwich Proposition: Enterprise Type

For the system in Fig. 2.6, the recent results are summarized in Fig. 2.7. The formulation in Fig. 2.7a was provided in [3], and the formulation in Fig. 2.7b was described in Sect. 2.1.3 (Fig. 2.1). Finally, the formulation in Fig. 2.7c, the *IDTCBG* logics, is a type of Matsui's ME based on the independent component analysis (ICA) [10], as shown in Fig. 2.8, and is partially coincident with the Chinese "Kishotenketsu" (strategy story in logics, *IDTC*).

The basic form of the ICA is referred to in Fig. 2.8. The respective ICA stages proceed from the Introduction ($S_1$, $S_2$), Development ($X$) and Turn ($Z$) to Conclusion ($Y$), and the $U$ and $V$ are rotation matrices. Then, the solution space, $Y$, is extended to $W$ as follows:

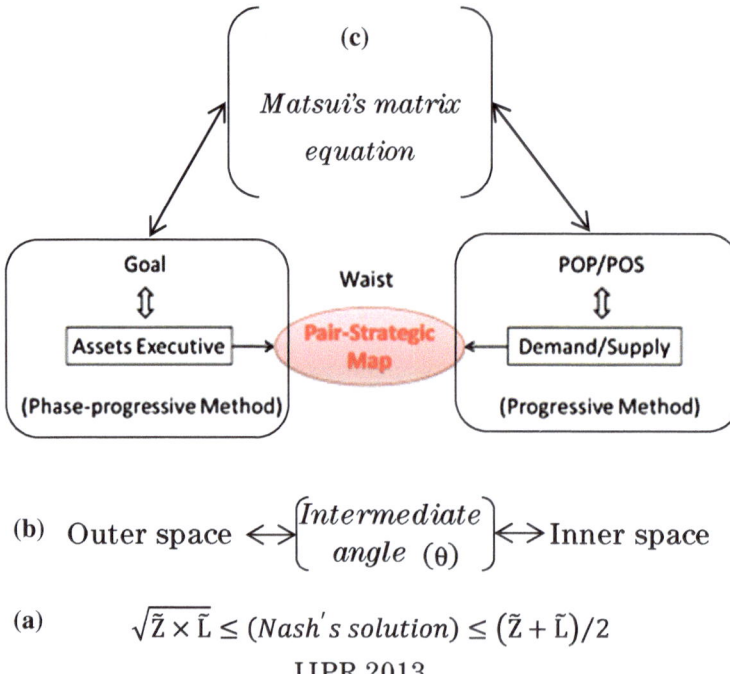

$$(\text{b}) \quad \text{Outer space} \leftrightarrow \begin{bmatrix} Intermediate \\ angle \ (\theta) \end{bmatrix} \leftrightarrow \text{Inner space}$$

$$(\text{a}) \qquad \sqrt{\tilde{Z} \times \tilde{L}} \leq (Nash's\ solution) \leq (\tilde{Z} + \tilde{L})/2$$

$$IJPR, 2013$$

**Fig. 2.7**  Three treatments of the sandwich model

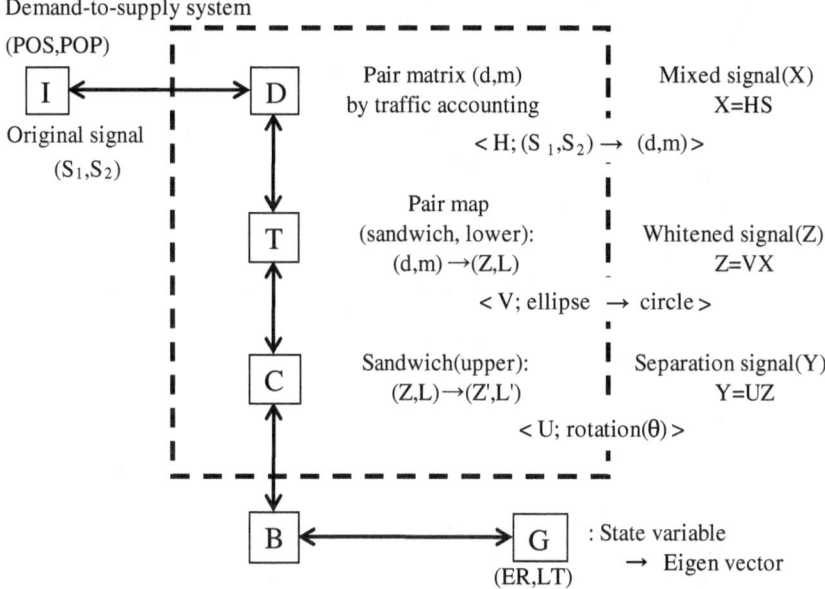

**Fig. 2.8** Matsui's matrix equation (Matsui's ME) with Independent component analysis (ICA). *I* Introduction, *D* Development, *T* Turn, *C* Conclusion, *B* Balancing, *G* Goal

$$W(=UTV) = H^{-1}, \qquad (2.18)$$

in which $T$ is the intermediate matrix of pair strategy (map) type.

These formulations have a common property; the rotation of the intermediate angle ($\theta$) is available as a strategic variable. This sandwich property of the enterprise type could support the sandwich proposition in nature versus artifacts body.

### 2.2.3    Type of Enterprise-Motor Body

The dualism of Matsui's, Newton's, and Jewell's equations can be visualized as shown in Fig. 1.3 This would support the sandwich proposition of an electric body. This example is provided for the electric motor of PMSM type in Fig. 2.9 [11, 12], and is similar to the sandwich property shown in Fig. 2.7.

This example is an engineering type of an artifact, and is composed of two rotations: Clark ($\alpha\beta$) and Park ($dq$) transformations. The formulation is similar to the type of Matsui's ME in Fig. 2.10. These findings could provide the clues to the auto-control enterprises, and might enable the realization of the enterprise-motor robot (body) of analog type.

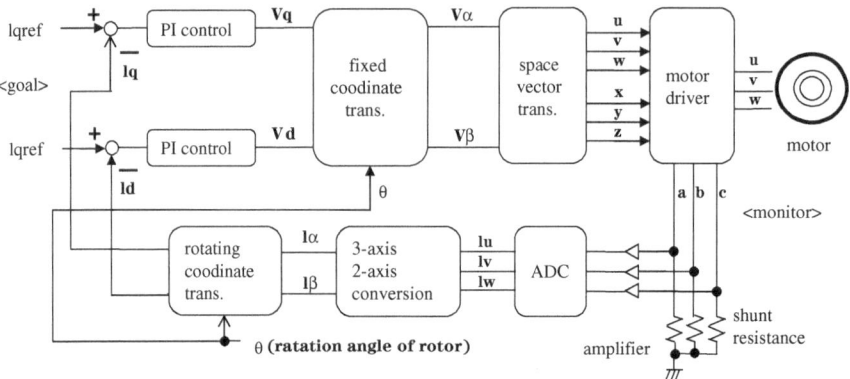

**Fig. 2.9**  Vector control motor of PMSM type

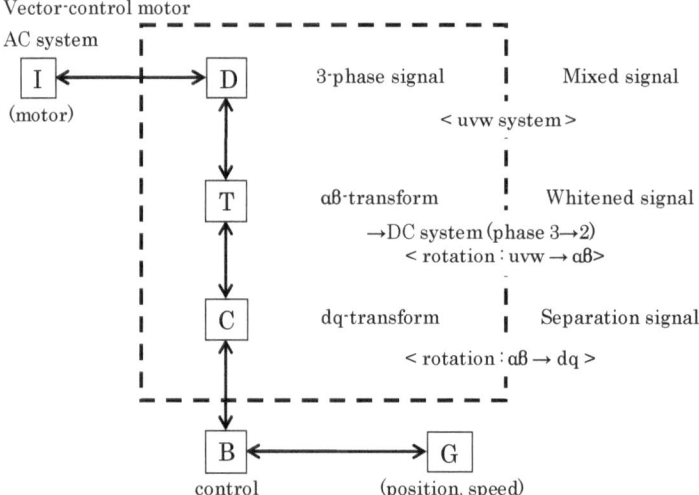

**Fig. 2.10**  Vector control motor with Matsui's ME

## 2.2.4   Type of Particle Physics Bodies

The type of particle physics bodies corresponds to the 3M&I-body in enterprise, as shown in Fig. 2.11. In this figure, the weight in Higgs denotes the amount of value, and the force in Graviton means the lead time in enterprise.

This formulation of a particle physics body is imaged as a type of Matsui's ME, and is proposed in Eq. (2.19) or Fig. 2.12, Chap. 2. In Fig. 2.11, the development (D) is the stage from the 3M&I to particle life cycle, and the conclusion (C) is the stage from the particle life cycle to pair map.

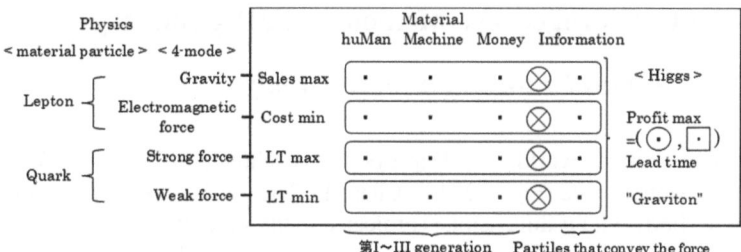

**Fig. 2.11**  3M&I-body of Particle physics type. Higgs boson: a particle that gives weight to the substance, graviton: a particle that mediates the gravitational interaction

| I |

$$(M_1 M_2 M_3 I) \times \begin{pmatrix} M_1, I_n \\ M_2, D \\ M_3, M_t \\ I, D_c \end{pmatrix}$$

| D |

| T |

$$\times (TM) \times \begin{pmatrix} I, LT_{max} \\ D, EC_{min} \\ M_t, LT_{min} \\ D_c, ER_{max} \end{pmatrix} \times$$

| C |

| B |  | G |

$$\begin{pmatrix} H_1 & G_1 \\ H_2 & G_2 \\ H_3 & G_3 \\ H_4 & G_4 \end{pmatrix} = (H \quad G)$$

M1,M2,M3:3M, I:Information, In:Introduction, D:Development,

Mt:maturity,   Dc:Decline,    B:Balancing,     G:Goal

H,H1,H2,H3:Higgs state <EN>, G, G1,G2,G3:Gravition state <LT>

$$(2.19)$$

**Fig. 2.12**  Type of 3M&I-body with Matsui's ME (2.19)

Generally, the type of Matsui's ME could be transformed into the corresponding characteristic equation as follows: For any 3M&I-body, the generalized equation is given by

$$Ax = \Lambda x \qquad (2.20)$$

where A is an operator, x is the proper state and $\Lambda$ is the proper value.

In Figs. 2.8, 2.10 and 2.12, the *IDTC* states are represented by

$$Ax_i = \lambda_i x_i, \quad i = 1, 2, 3 \qquad (2.21)$$

where $i = 1(I)$, $2(D)$ and $3(T)$. Then, Eq. (2.20) is composed of those of (2.21), and is as follows:

$$A = \begin{pmatrix} D & & 0 \\ & T & \\ 0 & & C \end{pmatrix}, \quad x = \begin{pmatrix} x_1 \\ x_2 \\ x_3 \end{pmatrix}, \quad \Lambda = \begin{pmatrix} \lambda_1 & 0 \\ 0 & \lambda_2 \end{pmatrix} \qquad (2.22)$$

## 2.3   Control: Progressive Method and 3D-Geometry

### 2.3.1   3D-Control Problem of 3M&I-Body

The administrative world of 3M&I-body is shown in Fig. 2.13 for the multidimensional space (Fig. 1.2 [13], Chap. 1). Then, it is suggested that the lower level of the body with $p$-dimension is projected to the superplane of the pair map and the upper level of the body is the vector space in the vertical axis of the superplane.

Usually, the $p$-dimension control might be not feasible. By using Matsui's equation, the $p/2$-vector control of the superplane could be feasible, as shown in Fig. 2.14. There, the $p/2$ denotes the contraction from $p$ to the two dimensions of vectors, $Z$ and $L$. By using this contraction, the 3D control of 3M&I-body could be made feasible under the strategy $(\theta, \varphi)$.

For example, let the $x_i$'s be denoted as $Z$ and $L$ in $W = ZL$. Then, the strategic $(\theta, \varphi)$ is from the undertarget goal $(AH)$ (1.1) as follows.

$$\exists (a^*|\theta, \varphi) \text{ s.t. } \min f(a) = AH(\theta, \varphi), \tag{2.23}$$

in which $f(a)$ is the function of least squares:

$$f(a) = ||Fa - y||^2. \tag{2.24}$$

**Fig. 2.13** 3D diagram: superplane (pair map) and its vector space

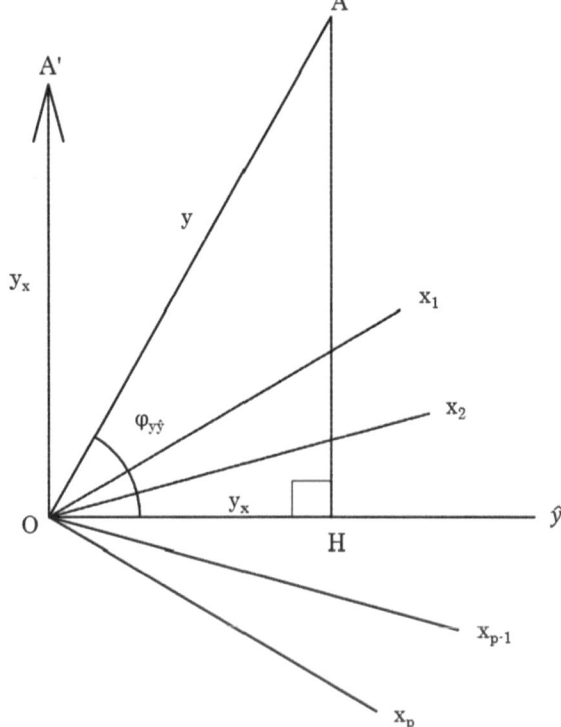

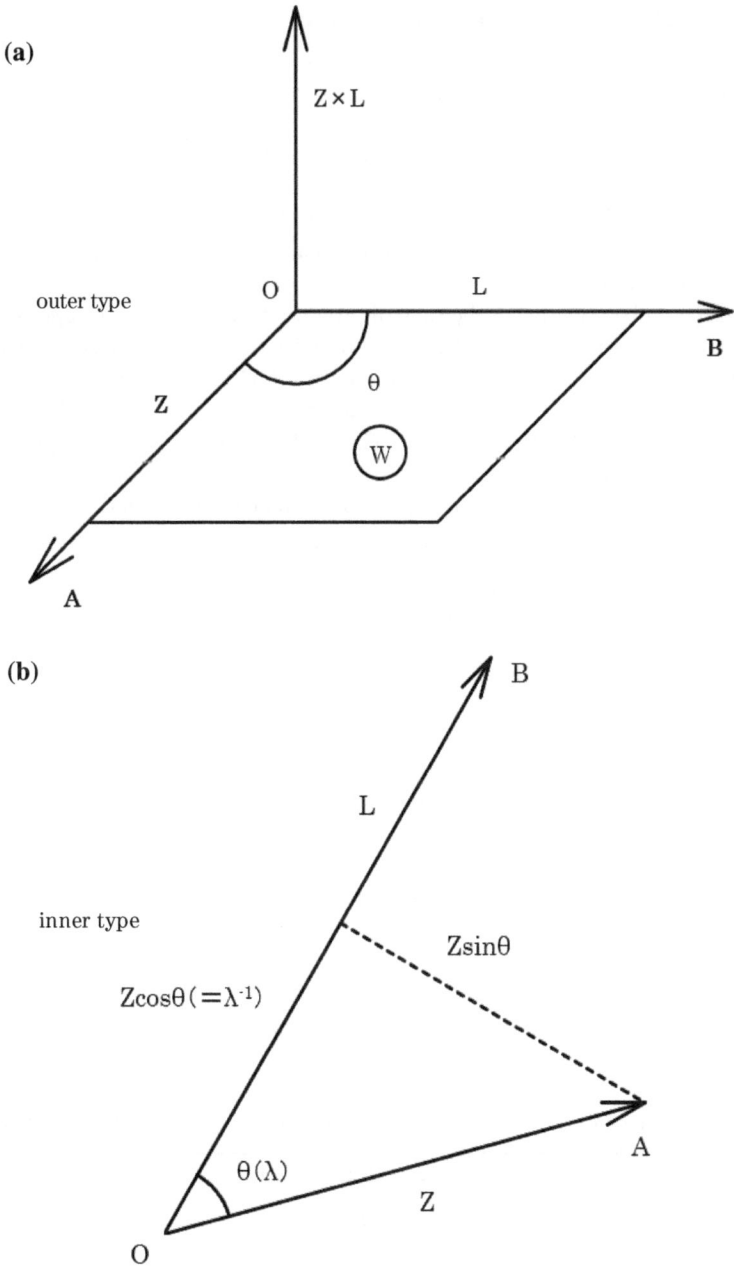

**Fig. 2.14**  Outer (**a**) versus inner (**b**) product at Matsui's equation

## 2.3.2  Progressive and Matsui's Logic

Generally, the equation of progressive control is from [4] as follows:

$$O_{t+1} = \overline{D}_{t+1} + MI_{t+1} - I_t, \quad t = 0, 1, 2, \ldots \tag{2.25}$$

where $O_t$ is the next order (supply) quantity, $\overline{D}_{t+1}$ is the look-ahead demand indicator, $MI_t$ is the moving-standard (marginal) inventory, and $I_t$ is the present amount of inventory.

The $MI_t$ can be regarded as the state of progressive control, and is represented as the node of the network in Fig. 2.15. This progressive method is also available for 3D control, similar to the new inventory control, ODICS, as in [4].

For ODICS, Matsui's logic [4] is useful and effective in the on-demand progressive control of the look-ahead type. This will be also applied to the progressive control of $p/2$ type in 3D.

Now, let us denote the demand speed and Chameleon's criteria by $Z(>0)$ and $\overline{\beta}(0 < \overline{\beta} < 1)$. If the following relation is satisfied by the principle of the lever (1.3):

$$Z_t \overline{\beta}_t = Z_{t+1} \overline{\beta}_{t+1}, \quad t = 0, 1, 2, \cdots \tag{2.26}$$

The $(t+1)$-epoch Chameleon's criteria, $\overline{\beta}_{t+1}$ is given by

$$\overline{\beta}_{t+1} = (Z_t / Z_{t+1}) \overline{\beta}_t. \tag{2.27}$$

Also, Matsui's logic of the second order is obtained for acceleration as follows:

$$\overline{\beta}_{t+1} = (Z_{t+1} / Z_{t-1}) \overline{\beta}_{t-1}. \tag{2.28}$$

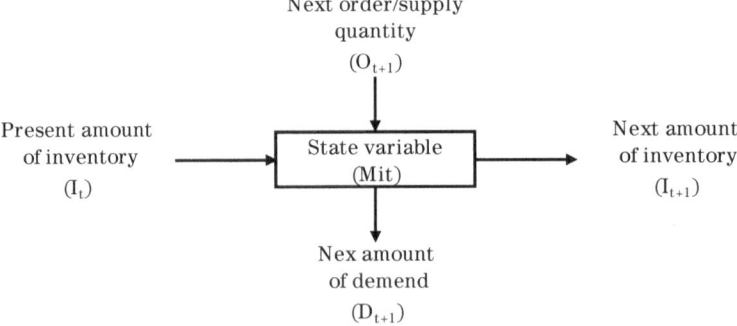

**Fig. 2.15** State variable of progressive control: $MT_{t+1} = O_{t+1} + I_t - \overline{D}_{t+1} \rightarrow I_{t+1}$

From (2.26) and (2.28), the second order $\overline{\beta}_{t+1}$ becomes

$$\overline{\beta}_{t+1} = (Z_t Z_{t+1} / Z_{t-1}^2)\overline{\beta}_t. \tag{2.29}$$

### 2.3.3   Medium Estimation of State

Generally, the medium criterion is important and is satisfied to the newsvendor [14] or revenue control types [15] in IEOR. Here, we consider the medium estimation of state.

Now, the inequality for the medium criterion, Y, is as follows:

$$(X - N)^+ > Y > (N - X)^+, \tag{2.30}$$

where $N$ is the number of units in demand or no-shows, and $X$ is the units ordered or overbooked.

This stochastic inequality, $P(X > N)$, becomes the trade-off (medium) of the underestimate cost and overestimate cost of demand in Fig. 2.16. In this figure, the penalties (costs) of the three types are introduced as follows:

$\beta_1(C_h)$   : maintenance cost of criterion
$\beta_2(C_o)$   : cost of overestimating demand or no-shows (opportunity cost)
$\beta_3(\rho)$   : cost of underestimating demand or no-shows (price)

By Fig. 2.16, the following inequality can be expressed:

$$P(X < N)p \geq P(X > N)C_o + C_h. \tag{2.31}$$

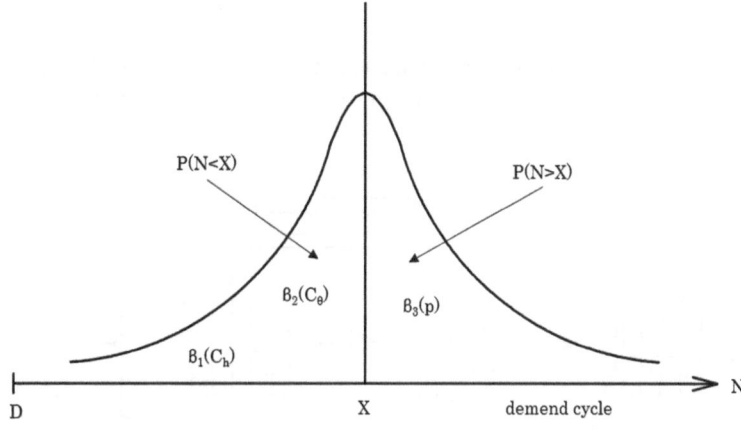

**Fig. 2.16** Newsvendor versus revenue control

For revenue control type, and for newsvendor type, the inequality is obtained:

$$\beta_1 + \beta_2 P(N < X) \geq \beta_3 P(N > X). \tag{2.32}$$

Thus, the lower and upper bound for P(X > N) is given by

$$\frac{p - C_h}{p + C_o} \geq P(X > N) \geq \frac{\beta_3 - \beta_1}{\beta_2 + \beta_3}. \tag{2.33}$$

The inequality (2.32) shows the medium estimation of state.

### 2.3.4  3D Cell and Graphics Modeling

For the control of 3M&I-body, we should administer the sandwich body in real time in a sustainable manner as possible. Then, the effective 3D graphics would be transferred to the $p/2$ control problem, and this would become a comprehensive and visual method to the Matsui's ME.

The cell modeling is well known at the cell automaton model. Figure 2.17 shows the one- and two-dimensional discrete conveyor types [16]. These are the scalar type with $Z$ or $Z \times Z$, and can be extended to the type of 3D modeling by a 3D cell, $Z \times Z \times Z$.

Then, the 3D cell in Matsui's equation can be visualized in the ($Z$, $L$, $W$)-space, and is shown in Fig. 2.18. In the ($Z$, $L$, $W$)-space, as shown in Fig. 2.18, the unit of the 3D cell is $Z \times Z \times Z$. For example, $Z_i'$s are accounts, $L_i'$s are dates, and $W_i'$s are amounts ($Z$, $L$) with different units, $Z'$s.

### 2.3.5  Two Types of 3D Dynamism

The two types of 3D modeling are an accounting system (lower level in sandwich), as shown in Fig. 2.18, and finance system (upper level in sandwich), as shown in

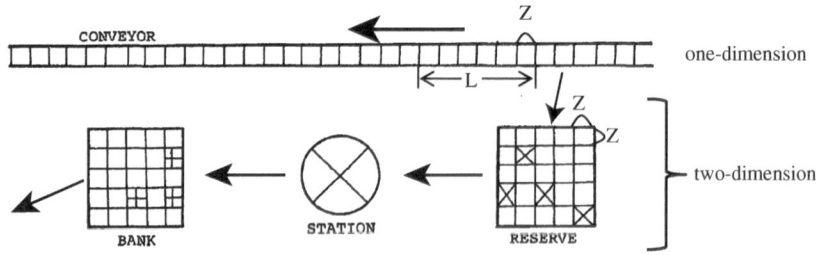

**Fig. 2.17**  One- and two-dimensional cells in discrete conveyor type

**Fig. 2.18** 3D cell in Matsui's
equation: W (value) = Z
(unit) × L (amount)

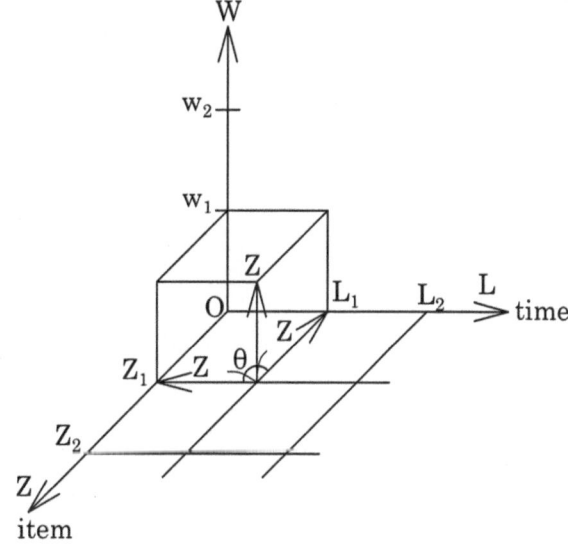

Fig. 2.19. The former is based on progressive figures, and the latter is based on the progressive phase in (2.34) and (2.35), respectively.

$$\text{position } (d, m) \rightarrow \text{progressive figure } (Z, L), \qquad (2.34)$$

$$\text{plane } (Z, L) \rightarrow \text{progressive phase } (\overline{W}, \overline{L}). \qquad (2.35)$$

**Fig. 2.19** Dynamism of the
3D accounting system at the
lower level

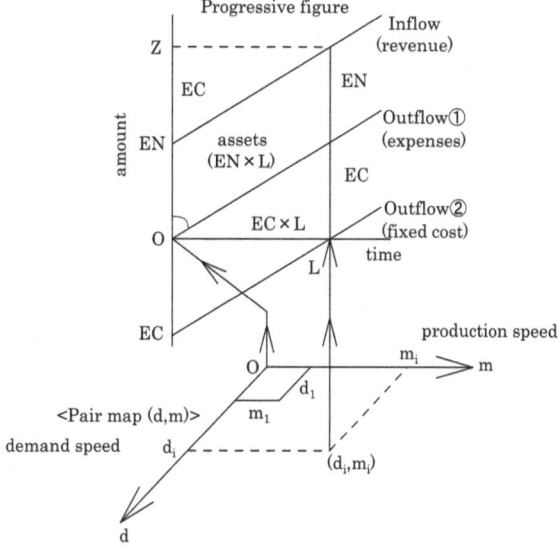

**Fig. 2.20** Dynamism of 3D finance system at the upper level

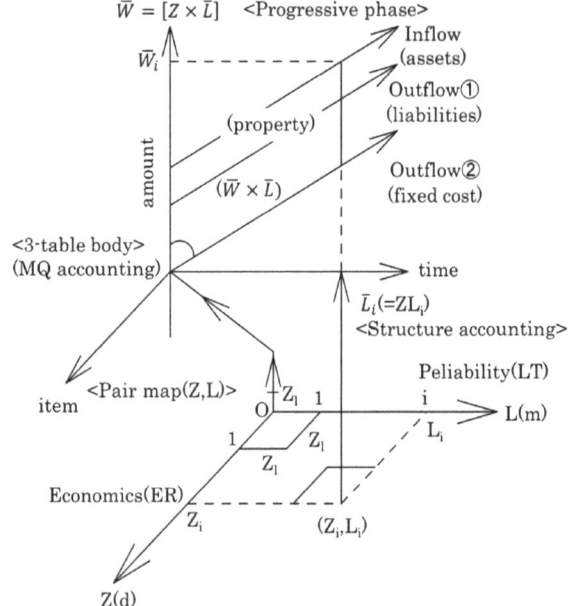

This 3D dynamism modeling would be useful and effective in the 3D progressive control of enterprises. $\overline{W}$ is represented by $\overline{W} = Z\overline{L} = Z^2L$ in Table 1.4. For example, when $Z$ is the unit revenue on the $Z$-axis and the width $Z$ on the $L$-axis is unit cycle, the fund $\overline{W}$ is (unit revenue) × (unit cycle) × (amount) in 3D.

Moreover, the sandwich control can be visualized in real time by the progressive types of both Figs. 2.19 and 2.20, and the multivariate management can be performed at $W = ZL$ by using the Eqs. (2.19) and (2.20), under any goal ($\overline{W}$). This is an alternative to the analytical method that uses Matsui's characteristic equation, and is the 3D geometric approach, to the sandwich management of enterprise dynamism.

# References

1. Matsui, M. (2008). *Manufacturing and service enterprise with risks: A stochastic management approach*. Springer.
2. Matsui, M. (1983). Game-theoretic consideration of order-selection and switch-over policy. In *Preprints of Japan Industrial Management Association, Fall Meeting* (pp. 48–49) (in Japanese).
3. Matsui, M. (2013). An enterprise-aided theory and logic for real-time management. *International Journal of Production Research*, 51(23,24), 7308–7312.
4. Matsui, M. (2014). *Manufacturing and service enterprise with risks II: The physics and economics of management*. Springer.

5. Matsui, M. (2013). Product × enterprise strategy: A matrix approach to enterprise systems for sustainability management, In *Proceedings of 14th Asia pacific Industrial Engineering and Management Systems Conference, Cebu, Philippines.*
6. Matsui, M. (2005). CSPS model: Look-ahead controls and physics. *International Journal of Production Research*, 43(10), 2001–2025.
7. Saaty, T. L. (1980). *The analytic hierarchical process: planning, priority setting, resource allocation.* McGraw-Hill.
8. Matsui, M. (2011). Conveyor-like network and balancing. In A. B. Savarese, *Manufacturing engineering*, Ch. 3 (pp. 65–87). NOVA.
9. Matsui, M. (2002). A management game model: Economic traffic leadtime and pricing settings. *Journal of Industrial Management Association, 52*, 1–9.
10. Hyvarinen, A., Karhunen, J., Oja, E. (2001). *Independent component analysis.* Wiley-Interscience.
11. Krishnan, K. (2010). *Permanent magnet synchronous and brushless DC motor drives.* NY: CRC Press.
12. Web, http://toshiba.semicon-storage.com/jp/sitemap.html (in Japanese).
13. Takeuchi, K., Yanai, H. (1972). *Foundations of multivariate analysis–projection method to linear space.* Tokyo: Touyoukeizai-Shinpousha (in Japanese).
14. Weeks, J. K. (1979) Optimizing planned lead time and delivery dates, In *21th Annual Conference Proceedings, APICS* (pp. 177–188).
15. Gross, R. G. (1997). *Revenue management.* Crown Business.
16. Beightler, C. S., Crisp, Jr., R. M. (1968). A discrete-time queueing analysis of conveyor-serviced production stations. *Journal of Operational Research*, 16(5), 986–1001.

# Chapter 3
# Science/Balancing of Multibody Systems

**Abstract** In general terms, the world of nature is physically compatible to a cyclic type, while the world of artifacts is restricted to that of a limited cycle because of limited resources (material, space, and time). In this environment, the sustainable recirculation and balancing problem could be regarded as the cooperative and synergistic use of limited resources (cycle). There are two types of deterministic and stochastic scheduling/balancing methods described in Sects. 3.1 and 3.2, respectively. In Sect. 3.1, we present the dual relation of a job shop (JS) and a line system (OE) by using the flow time approach in order to identify the scheme of Matsui's equation. This would elicit an optimal/balancing condition beyond Johnson's rule "bowl effect". Furthermore, in Sect. 3.2, we consider an optimal arrangement to minimize the total expected risk on limited-cycle scheduling, in accordance with the stochastic type. A recursive formula for the total expected risk is presented next, and its algorithm is proposed based on the Branch and Bound method. Finally, the so-called "bowl effect" is also refounded as the balancing property by introducing a theorem and conducting numerical experiments.

**Keywords** Scheduling · Balancing · Matsui's equation · Limited cycle · Johnson's rule · Bowl effect

## 3.1 Dual Job Shop Versus Line System and Balancing

### 3.1.1 Introduction to 3M&I Balancing

In the job shop (JS) domain, scheduling issues are well known as objects of major class type [1]. The so-called "line balancing problem" [2] is also similar. This study focuses on the corresponding relation for both classes in order to identify a dual relation for the JS versus the conveyor system from [3].

The problem has been considered using a flow time approach in station-centered issues [3]. In there, the so-called "Matsui's equation" [4, 5], whereby $W = ZL$ in

© The Author(s) 2016
M. Matsui, *Fundamentals and Principles of Artifacts Science*,
SpringerBriefs in Business, DOI 10.1007/978-981-10-0473-5_3

queuing, shows the relationship between waiting line and time. However, herein, this relation is reconsidered from the view of factory physics, as presented in 1995 [6], and factory science, as presented in 2005 [7]. That is, this dual system belongs to the same class as Matsui's equation, and it is assumed to have the same type in accordance with system balancing issues.

Section [8] focuses on the dual relation of JS and the conveyor system, and presents the major principle on space versus time in the production system. These issues show that it is necessary to reevaluate Matsui's equation, and suggest the optimal/balancing condition in the discrete science of a 3M&I system.

It has already been determined that Matsui's equation and the economic order quantity (EOQ) formula are equivalent in lot production [5]. These results would suggest the unification of Matsui's equation in the discrete world. In the near future, the expansion of Johnson's rule is expected, and the type of limited-cycle problem [9–11] should be reconsidered and developed furthermore.

Finally, it is noted that the proposed method would be effective in developing an optimal rule in the JS problem, namely, $n/k/Fmax$ $(k > 4)$, as the optimal rule would be nearer to the balancing principal.

## 3.1.2  Job Shop and Flow Time

### 3.1.2.1  Job Shop Problem and Example

Let us consider a simple example of a JS problem in Table 3.1 [12]. Table 3.1 lists the work times in a problem comprised of six jobs and one machine case. This classical scheduling problem is the sequencing class minimizing the mean flow time $\bar{F}$.

The mean flow time $\bar{F}$ is usually obtained using the following equation:

$$\bar{F} = \frac{1}{6} \sum C_i \tag{3.1}$$

**Table 3.1** Scheduling problem of n/1/$\bar{F}$ type [8]

| Job ($i$) | Work time ($x_i$) |
|-----------|-------------------|
| A | 7 (min) |
| B | 6 |
| C | 4 |
| D | 3 |
| E | 2 |
| F | 1 |

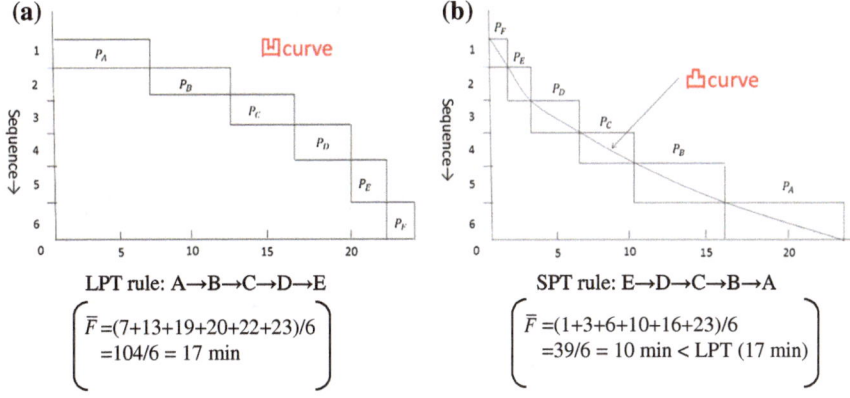

**Fig. 3.1** $n/1/\bar{F}$ solution: LPT versus SPT ($F$ mean flow time)

where $C_i$ is the completion time of $i$th job. It is noted here that $\bar{F}$ corresponds to $W$ in Matsui's equation, namely $W = ZL$.

For this $n/1/\bar{F}$ example there is the graphical solution depicted in Fig. 3.1. Figure 3.1a depicts the LTP solution, and Fig. 3.1b depicts the SPT solution.

### 3.1.2.2 Dual JS Versus Conveyor System

Generally, the job shop problem with multiple machines of flow type (FS) would correspond to a conveyor system of an ordered-entry type (OE) in accordance with a physical view. That is, both types become a part of the same class if the job number ($n$) corresponds to the station number ($K$).

$$\text{Job number}(n) \Leftrightarrow \text{Station number}(K) \qquad (3.2)$$

Figure 3.2 shows the conveyor system of an OE type with a production rate, $r$, and an overflow rate $v$. The sequencing problem in FS and the arrangement problem in OE correspond to the same system.

In addition, Fig. 3.3 shows an example of a two-machine/line case. Similar to Fig. 3.2, there exists a dual relation.

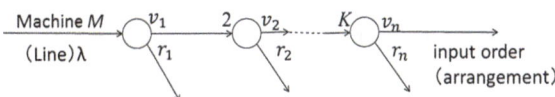

**Fig. 3.2** Correspondence of job shop (FS) versus the conveyor system (OE)

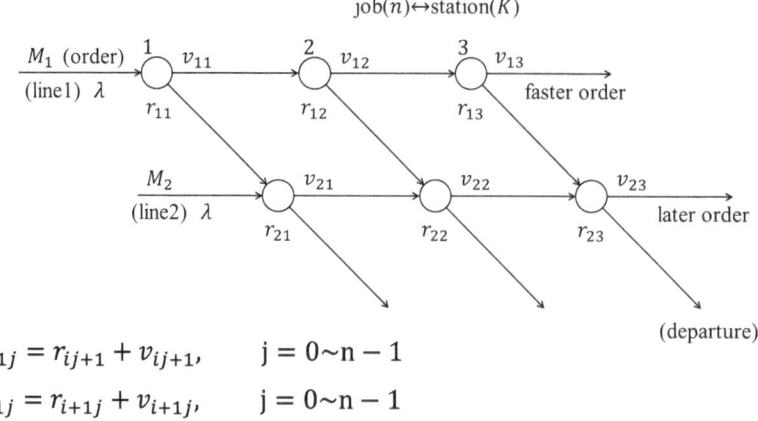

$$v_{1j} = r_{ij+1} + v_{ij+1}, \qquad j = 0\sim n-1$$
$$r_{1j} = r_{i+1j} + v_{i+1j}, \qquad j = 0\sim n-1$$

**Fig. 3.3** Dual case of a two-machine/line system

## 3.1.3  Flow Time Formula and Example

### 3.1.3.1  Single-Machine (Line) Case

Given the type of Fig. 3.2, the mean flow time is easily computed from the performance formula of the conveyor system [3, 9]. The following formulae mathematically express the respective types of line and ordered entry:

$$F_I = \sum_{i=1}^{K} W_i = \lambda^{-1} \sum Z_i L_i, \text{line type} \tag{3.3}$$

$$F_{II} = \left(\sum_{i=1}^{K} L_i\right)\big/r_{II} = \left(\sum_{i=1}^{K} L_i\right)\big/(\lambda - v_k), \text{OE type} \tag{3.4}$$

The formulae for $F_I$ and $F_{II}$ are now applied to the data of Table 3.1. Correspondingly, the computations are as follows:

$$\begin{aligned}
\text{Line}: \quad F_I &= \sum W_i = \sum Z_i L_i, \\
&\quad (z_i \to x_i, L_i \to n-i+1) \\
&= 1 \times 6 + 2 \times 5 + 3 \times 4 + 4 \times 3 + 6 \times 2 + 7 \times 1 \\
&= 59(104) \leftrightarrow FS(\text{Total flow time } W)
\end{aligned} \tag{3.5}$$

$$\begin{aligned}
\text{OE}: \quad F_{II} &= \left(\sum L_i\right)\big/(\lambda - v_k), \ (L_i \to C_i) \\
&= (1+3+6+10+16+23)/6 \\
&= 59/6 = 10(17) \\
&\leftrightarrow FS(\text{Mean flow time } \bar{F})
\end{aligned} \tag{3.6}$$

in which $\lambda = 1/6$, $K = n = 6$ and $v_{K(n)} = 0$. The traditional solution is also seen in Fig. 3.1b.

### 3.1.3.2 Two-Machines (Line) Case

Let us consider the case of an $n$-job, $m$-machine, and work time, $x_{ij}$, $i = 1 \sim n$, $j = 1 \sim m$, in FS. The formulae for line and OE types are summarized as follows:

$$\text{Line}: F_I = \sum Z_i L_i = \sum (n - i + 1)\bar{x}_i, \tag{3.7}$$

$$\text{OE}: F_{II} = \left(\sum L_i\right)/\lambda = \left(\sum C_i\right)/\lambda. \tag{3.8}$$

where the work time of the $i$th job is

$$\bar{x}_i = \sum_{j=1}^{m} x_{ij}, \quad i = 1, 2, \ldots, n. \tag{3.9}$$

Table 3.2 summarizes the scheduling problem of a two-machine case in [13] (see Fig. 3.3). This sequencing problem aims to estimate the maximum flow time $F_{max}$. From the Eqs. (3.7) and (3.8), the mean flow time is computed in accordance with Johnson's rule as follows:

$$\begin{aligned} F_{max} &= (7 + 17 + 32 + 47 + 52)/5 = 155/5 = 31 \leftrightarrow F_{II} \\ W &= 5 \times 7 + 4 \times 10 + 3 \times 15 + 2 \times 15 + 1 \times 5 = 155 \leftarrow F_I \end{aligned} \tag{3.10}$$

## 3.1.4 Duality Relation Using a Flow Approach

### 3.1.4.1 Flow Time and Matsui's Equation

From the results listed in Eqs. (3.5) and (3.6), the following relation between $F_I$ and $F_{II}$ is identified for the data of Table 3.1, as shown in Table 3.3.

| Table 3.2 Scheduling problem of a two-machine type [13] | Job number | Work time (day) | |
|---|---|---|---|
| | | Machine $M_1$ | Machine $M_2$ |
| | 1 | 3 | 2 |
| | 2 | 1 | 6 |
| | 3 | 8 | 7 |
| | 4 | 4 | 6 |
| | 5 | 11 | 4 |

**Table 3.3** Flow time and dual relation for Table 3.1

| Dual | $F_I = K(n) \times F_{II}$ | | | Dual equation (D) |
|---|---|---|---|---|
| Main | $W = L \times Z$ | | | Matsui's equation (M) |
| Table 3.1 | 59 | 6 | 10 | SPT rule case |
| | 104 | 6 | 17 | LPT rule case |

**Fig. 3.4** Johnson's rule solution and balancing

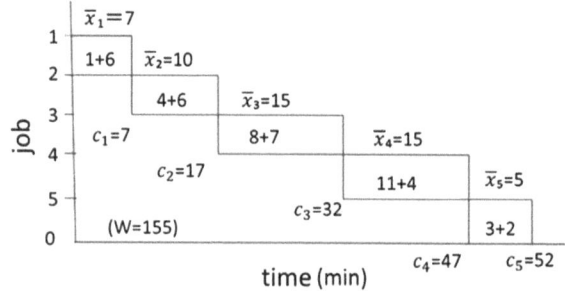

In Table 3.3, $W$ is the sojourn time, $Z$ is the cycle time, $L$ is the number of cycles, and $K(n)$ is the number of stations and jobs. Table 3.3 shows the dual relation of $F_I$ and $F_{II}$ in $D$, and this corresponds to M in Matsui's equation ($W = ZL$) [4, 5]. This duality relation in flow would become the major principle on the production system and in factory physics and science [8].

In Table 3.2, Johnson's rule gives the maximal flow time $F_{max}$. From Eq. (3.10), the following dual relation (3.11) is deduced:

$$\begin{aligned} F_I &= K(n) \times F_{II} \\ &= 5 \times 31 \\ &= 155 \leftrightarrow W = L \times Z : \text{ Matsui's equation} \end{aligned} \tag{3.11}$$

It is here that the Gantt chart yields $F_{max}(F_{II}) = 30$.

Figure 3.4 shows the solution of Johnson's rule in Table 3.2, and the problem is summarized in a convex curve (bowl effect), toward a straight line.

### 3.1.4.2   Duality and Balancing View

For the multiline (machine), there would be a dual relation between FS and OE, and this relation would be unified from the view of Matsui's equation. This unification is deduced in accordance to Eq. (3.12) as follows:

$$W = n\bar{F} = \sum C_i = \sum (n - i + 1)\bar{x}_i \tag{3.12}$$

1. Single line
   For one line (OE) versus the SPT rule (FCFS),

$$F_I = K(n) \times F_{II} \leftrightarrow W = L \times Z : \text{ Matsui's equation} \qquad (3.13)$$

2. Two-line (FS)
   For two-line (OE) versus Johnson's rule,

$$F_I = K(n) \times F_{II} \leftrightarrow W = L \times Z : \text{ Matsui's equation} \qquad (3.14)$$

3. Multiline

$$\text{For a multiline (OE) versus a } k\text{-out-of-}n \text{ system} \\ \leftrightarrow \text{Matsui's equation}(W = L \times Z) \qquad (3.15)$$

In Eq. (3.15), $k$ and $n$ are the respective number of machines and jobs. Specifically, when $n > k$, the system is an FS type, and when $n < k$, it is a mixed line type.

Moreover, it is remarked from $W$ in Eqs. (3.5) and (3.10) that the following balancing principle is supposed and conjectured:

$$W_1 = W_2 = \cdots = W_n \quad \text{for each job.} \qquad (3.16)$$

Because the classical inequality and Matsui's equation results in the following inequality of additional versus an additive mean:

$$\sum (Z_i L_i)/6 \geq \sqrt[6]{\prod (Z_i L_i)} \qquad (3.17)$$

The equality condition would then result in the optimal equation:

$$Z_i L_i = Z_j L_j, \quad i \neq j. \qquad (3.18)$$

Thus, complete balancing would show a straight line response instead of a convex curve in Fig. 3.4.

Based on Eqs. (3.16)–(3.18), the objective for optimality/balancing would be formulated as follows:

$$\sum Z_i L_i \rightarrow \min \qquad (3.19)$$

It is remarked that this objective also applies in [5] and [14].

## 3.2  Bowl Balancing and Arrangement of Bodies

### 3.2.1  Limited-Cycle Problem with Multiperiod

Scheduling problems are concerned with the allocation of limited resources over time for a manufacturing period and multiple other periods [10, 13]. The allocation also affects the optimality of a schedule with respect to various criteria such as production cost, idle risk, and delay risk. Under the condition of uncertainty, the result or efficiency of a period is often controlled not only by the risks of this period but also by the risks generated beforehand.

Whether the process (period or seat) satisfies the due time (restriction) or not is usually dependent on the state of the past process, as seen in [15–17]. In particular, in the case of the risk that depends on the situation of a past process (for instance, the case of a manufacture line for a multiperiod), the choice of the most efficient and economical assignment (optimal assignment) of machines, workers, or jobs, is an important problem in load/risk planning (see, for example [18, 19]).

A limited-cycle scheduling class presents the following problems; Verzijl [16] analyzed the element and construction of a production system [20] and presented a framework for the analysis of delays within this system [15]. It also provided a review for the origin and solution of a period batch control system [21] concerned with the problem of scheduling a set of jobs associated with random due dates on a single machine in order to minimize the expected maximum lateness in a stochastic environment.

We consider herein a "limited-cycle model with multiple periods" [22]. In this model, there exists an object with some constraints. If the object does not satisfy these constraints, this becomes a risk, and the (object) cycle occurs repeatedly over multiple periods (or processes). This model is applied to manufacturing lines, time-bucket balancing, production seat systems (Fig. 3.5) and to other applications [9, 22, 23]. In the case of manufacturing lines, important considerations include the production time as the object, the due time of production time as the constraints, and the idle or delay time as the risk.

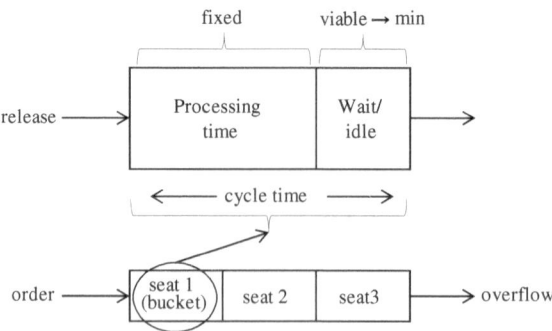

**Fig. 3.5** Production seat model

In this section, we consider an optimal arrangement in order to minimize the total expected risk in limited-cycle problems with multiple periods. First, we systematically classify and model the multiperiod problem. Next, we present the recursive formula for the total expected risk and propose an algorithm for optimal assignments aimed at the minimum total expected risk in limited-cycle problems with multiple periods, based on the Branch and Bound method.

By investigating the execution time with a computer, we researched the optimal assignments by using numerical experiments. The results obtained in this section are useful for the design and the production of a seat system. As future work, we will investigate the problem of optimal assignments in detail. For example, the property of optimal arrangement in which one or several workers has special processing efficiency.

### 3.2.2   Multi Limited-Cycle Modeling

#### 3.2.2.1   Stochastic Approach to Multicycle

The stochastic approach to multiple cycles [3, 24] is seen in Fig. 3.6. In Fig. 3.6, the multicycle and interference is related to the stop rate as follows:

$$\text{Stop rate} = \frac{1}{Z} \prod_{i=1}^{n} P_r(T_i > Z), \tag{3.20}$$

in which the target time (cycle time), $Z$, corresponds to A(7), B(6), C(4), D(3), E(2), and F(1) in Table 3.1.

For example, the behavior of the stop rate is shown in Fig. 3.7 for the exponential service with processing rates $\mu = 1.0$ and 1.2. In Fig. 3.7, the optimal cycle time $Z^*$ becomes $Z^* = 2.0$ for both processing rates.

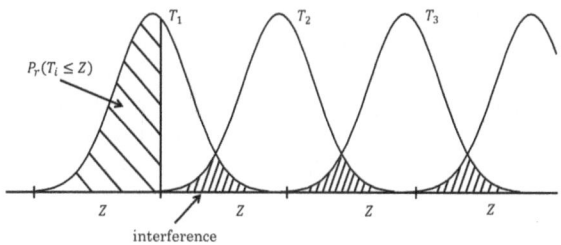

**Fig. 3.6** Multiple limited-cycle ($Z$) and interference

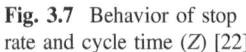

**Fig. 3.7** Behavior of stop rate and cycle time (Z) [22]

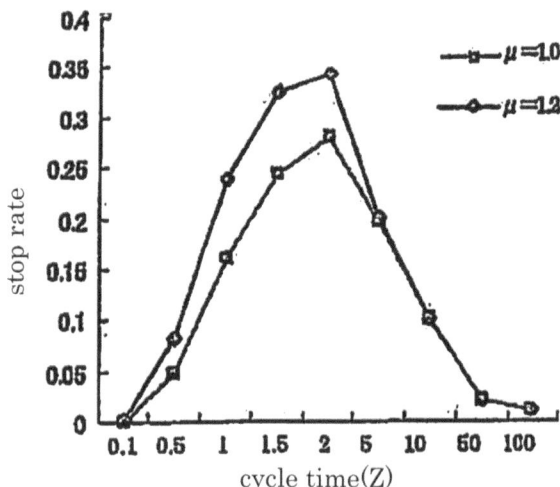

### 3.2.2.2  Multilimited Cycle Model with Penalties

Before we describe the multilimited cycle model in detail as an example, we consider a manufacturing line, as shown in Fig. 3.9 [11]. The manufacturing line has $n$ processes, and each process is executed at one period. A product is produced through the manufacture line. First, a material is processed at process 1 ($P_1$, whereby the process is executed at period 1, while the processed material is at process 2 ($P_2$, period 2), at process 3 ($P_3$, period 3), and in turn, at process $n$ ($P_n$, period $n$) [24].

We now consider the following situation. The limited production time (Z) such as the target production time or cycle time, is given for $n$ periods. In this situation, a delay or idle time occurs in each period. When the delay occurs, this usually influences the following periods. We describe this effect not temporally but financially in terms of costs. This is because the delay can be recovered by the overtime work or spare workers in each period. With regard to the delay cost, we consider the cost per unit time ($C_l^{(k)}$), which occurs within the period, given that the delay occurs in $k$ consecutive periods before the specific period.

When idle occurs, we consider the cost per unit time ($C_s$) because stocks of the products will be generated. Furthermore, we also consider the cost per unit time ($C_t$) for the limited production time (Z) as a fixed cost. Figure 3.8 shows the above costs in the model. In Fig. 3.9, $C_s$ occurs owing to the idle at period 1, $C_l^{(1)}$ occurs owing to the delay at period 2 ($i$) since the idle occurs at period 1 ($i - 1$), and $C_l^{(2)}$ occurs

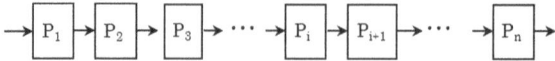

**Fig. 3.8** A manufacture line with the probability of processing

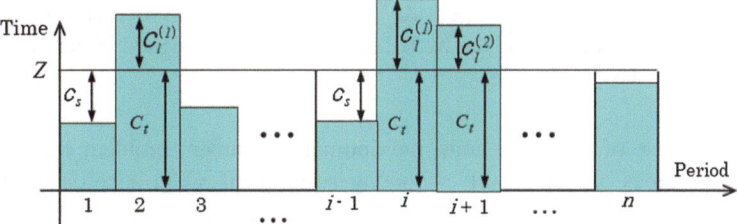

**Fig. 3.9** Costs in the limited-cycle model with penalties

owing to the delay at period $i + 1$ since the idle occurs at period $i - 1$ and the delay occurs at period $i$.

Based on the above situation, we consider the model of a multilimited cycle model, referred to [25–27], under the following assumptions:

(1) The number of periods $n$ (it may be considered that $n$ is the number of production seats or production processes).
(2) The prescribed limited production time (or target production time or cycle time) is denoted by $Z$.
(3) The production time of one period is denoted by $T$ and is assumed to be exponentially distributed. The production times are assumed to be statistically independent. We then consider the following costs in the multilimited cycle model, as shown in [11].
(4) The cost per unit time ($C_t$) for the production time limit ($Z$) occurs in each period.
(5) When $T < Z$, the idle cost per unit time ($C_s$) occurs in each period.
(6) When $T > Z$, the delay cost per unit time ($C_l^{(k)}$) occurs within the period if the delay (that is, $T > Z$) occurs in $k$ consecutive periods before the specific period, for $k = 1, 2, \ldots, n$, where $C_l^{(1)} \le C_l^{(2)} \le \cdots \le C_l^{(n)}$.

We now consider the situation in which one worker is assigned to each period in the multilimited cycle model presented above. One of the most important problems is how to assign workers to periods for minimizing the expected cost over $n$ periods. We call such a problem the optimal assignment problem.

For stating the optimal assignment problem, we define the following notations:

$W$  the set $\{\mu_1, \mu_2, \ldots, \mu_n\}$, where $\mu_i$ is the mean processing rates of worker $i$ ($i = 1, 2, \ldots, n$), and $\mu_1 \le \mu_2 \cdots \le \mu_n$.

$\pi$  the permutation composed from 1 to $n$, where $\pi = (\pi(1), \pi(2), \ldots, \pi(n))$.

$S_n$  the set of $\pi$, where $\pi(i)$ is the number of workers assigned in period $i$. For example, when $n = 5$, $\pi = (2, 4, 1, 5, 3)$ is the case in which the workers with $\mu_2$, $\mu_4$, $\mu_1$, $\mu_5$ and $\mu_3$ are assigned in periods 1–5, respectively. In this paper, the permutation $\pi$ is called the assignment $\pi$ because $\pi$ denotes the assignment of workers.

*TC* (*n*; π, *W*)   the total cost in periods 1–*n* when the mean processing rates of the workers are given by *W*, and workers are assigned in periods 1–*n* by assignment π.

With the use of these notations, the optimal assignment problem with multiple periods becomes a problem of estimating assignment $\pi^*$ in accordance the following equation [28]:

$$TC(n; \pi^*, W) = \min_{\pi \in S_n} TC(n; \pi, W). \qquad (3.21)$$

## 3.2.3  Formulation and Theorem

### 3.2.3.1  Notation and Expected Cost

First, some notations are defined. For $l = 1, 2, \ldots, n$, $T_i$ is the production time of the worker with mean processing rate $\mu_l$. for $i = 1, 2, \ldots, n$, $C\left(i; T_{\pi(1)}, T_{\pi(2)}, \ldots, T_{\pi(i)}\right)$ is the idle or delay cost occurring in period $i$, where the workers with mean processing rates $\mu_{\pi(1)}, \mu_{\pi(2)}, \ldots, \mu_{\pi(i)}$ are assigned to periods 1–*i*, respectively.

Based on the assumptions (1)–(6) referred to in Sect. 3.2.2.2,

$$C(i; T_{\pi(1)}, T_{\pi(2)}, \ldots, T_{\pi(i)}) = \begin{cases} C_s \cdot (Z - T_{\pi(i)}) & \text{if } Z \geq T_{\pi(i)}, \\ C_l^{(i-j)} \cdot (T_{\pi(i)} - Z) & \text{if } Z \geq T_{\pi(j)}, Z < T_{\pi(j+1)}, \ldots, Z < T_{\pi(i)}, \end{cases}$$

$$(3.22)$$

where $j = 0, 1, 2, \ldots, i - 1$ and $j \equiv 0$ correspond to the case of lack of idle times (no idle times) within periods 1–*i*.

The expected cost $TC(n; \pi, W)$ in *n* periods when the workers are assigned according to assignment π is expressed as

$$TC(n; \pi, W) = nC_t Z + f(n; \pi(1), \pi(2), \ldots, \pi(n)), \qquad (3.23)$$

where

$$f(i; \pi(1), \pi(2), \ldots, \pi(i)) = E\left[\sum_{m=1}^{i} C\left(m; T_{\pi(1)}, T_{\pi(2)}, \ldots, T_{\pi(m)}\right)\right] \qquad (3.24)$$

for $i = 1, 2, \ldots, n$.

The expected cost $TC(n; \pi, W)$ is given for a given assignment π. However, the purpose of this paper is to obtain an optimal assignment. For this, we must obtain the expected costs for all assignments and identify their minimum expected costs.

For efficient execution of these procedures we use recursive formulas for the expected cost $TC(n; \pi, W)$ and the Branch and Bound method (this concept is described in [11]).

### 3.2.3.2   Theorem and Property

For obtaining the expected cost $TC(n; \pi, W)$ efficiently, useful recursive formulas are given by Theorem 3.1, which is provided in the next page. Some additional notations are defined below. For $l = 1, 2, ..., n$,

$P_l$   is the probability of worker $l$ with processing rate $\mu_l$ becoming idle, that is, $\Pr\{T_l \leq Z\}$.

$Q_l$   is the probability of the worker $l$ with processing rate $\mu_l$ becoming delayed, that is, $\Pr\{T_l > Z\}$.

$TS_l$   is the expected idle cost of the worker $l$ with processing rate $\mu_l$, and is defined as $E[Z - T_1|T_l \leq Z]Pr(T_l \leq Z)$.

$TL_l$   is the expected delay cost of the worker $l$ with processing rate $\mu_l$, and is defined as $E[T_l - Z|T_l > Z]Pr(T_l > Z)$.

Since the production times are exponentially distributed, it follows that,

$$P_l = \int_0^Z \mu_l e^{-\mu_l x} dx = 1 - e^{-\mu_l Z} \qquad (3.25)$$

$$Q_l = \int_Z^\infty \mu_l e^{-\mu_l x} dx = e^{-\mu_l Z} (= 1 - P_l) \qquad (3.26)$$

$$TS_l = \int_0^Z (Z - x)\mu_l e^{-\mu_l x} dx = Z - \frac{1}{\mu_l}\left(1 - e^{-\mu_l Z}\right) \qquad (3.27)$$

$$TL_l = \int_Z^\infty (x - Z)\mu_l e^{-\mu_l x} dx = \frac{1}{\mu_l} e^{-\mu_l Z}. \qquad (3.28)$$

**Theorem 3.1** *The expected cost $TC(n; \pi, W)$ can be obtained by Eq. (3.23) and the recursive formulae for $f(n; \pi(1), \pi(2), ..., \pi(n))$ are given by the following Eq. (3.29):*

$$f(i; \pi(1), \pi(2), \ldots, \pi(i)) = \begin{cases} f(i-1; \pi(1), \pi(2), \ldots, \pi(i-1)) \\ + \sum_{j=0}^{i-1} C_l^{(i-j)} \cdot TL_{\pi(i)} \ h(i-1, j) + C_s \cdot TS_{\pi(i)}, & \text{if } i > 1 \quad (3.29) \\ C_s \cdot TS_{\pi(1)} + C_l^{(1)} \cdot TL_{\pi(1)}, & \text{if } i = 1 \end{cases}$$

where $h(i, j)$ are here omitted [11].

Then, for $f(i; \pi(1), \pi(2), \ldots, \pi(i))$, the following property (Property 3.1) holds.

**Property 3.1** *For* $i = 2, 3, \ldots, n$ *and any* $\boldsymbol{\pi} \in S_n$,

$$f(i; \pi(1), \pi(2), \ldots, \pi(i)) \geq f(i-1; \pi(1), \pi(2), \ldots, \pi(i-1)). \qquad (3.30)$$

*Proof* Since all the costs occurred in period $i$, and are more than or equal to 0, Property 3.1 can be proved. □

### 3.2.4   Numerical Example for Assignment

#### 3.2.4.1   Case of Fixed Target Production Times

Now, we consider the expected cost of each assignment and the property of optimal assignment under fixed target production times (cycle times) [11]. We consider the case in which the numbers $n$ of periods are from 2 to 10, where $b = 20$ and the other parameters are the same as those in [23].

Table 3.4 shows the expected cost and optimal assignment in ascending and descending orders, respectively, in an assignment when the target production time ($Z$) is 2.0. In Table 3.4, we show the expected cost of the ascending order is the case

**Table 3.4** The expected cost and optimal assignment

| Production period ($n$) | The expectation cost | | | Optimal assignment |
|---|---|---|---|---|
| | Ascending order | Descending order | Optimal value | |
| 2 | 208.81 | 223.87 | 208.81 | (1,2) |
| 3 | 268.11 | 291.47 | 263.54 | (1,3,2) |
| 4 | 318.39 | 344.31 | 308.95 | (1,4,3,2) |
| 5 | 365.99 | 392.47 | 352.02 | (1,4,3,5,2) |
| 6 | 413.37 | 439.96 | 395.82 | (1,5,4,3,6,2) |
| 7 | 461.35 | 487.96 | 441.13 | (1,6,4,5,3,7,2) |
| 8 | 510.16 | 536.77 | 488.07 | (1,7,4,5,6,3,8,2) |
| 9 | 559.82 | 586.43 | 536.37 | (1,8,4,6,5,7,3,9,2) |
| 10 | 610.28 | 636.90 | 585.91 | (1,9,4,7,6,5,8,3,10,2) |

when workers are assigned in an order from a low to high processing rate, and the ending order is the case when workers are assigned in an order from a high to a low processing rate.

From Table 3.4, we can note that the expected cost of the descending order is higher than that of the ascending order. This is because the delay cost is considerably larger than the idle cost in this case. From Table 3.3, we also can note that the optimal assignment is given in accordance to $\pi = (1, n - 1, 4, n - 3, \ldots, n - 2, 3, n, 2)$, and shows a class of bowl effect [25].

### 3.2.4.2   Case of Viable and Inappropriate Target Production Times

Next, we consider the relation between the number of periods and the optimal target production time (cycle time) [11]. The target production time $(Z)$ is changed as shown in Table 3.5.

Table 3.6 shows the optimal target production time for the ascending order, descending order, and optimal value. In this case, the optimal value refers to the target production time of the optimal assignment with the least cost.

From Table 3.6, it can be noted that the target production time required to minimize the expected cost becomes small with the increase in the number of periods. This is because the idle cost and set cost of the target production time are higher than those of the delay cost when the number of processes increases.

In this section, we discuss the optimal assignment of the case in which one worker is in the system possessing an inappropriate (or good) processing rate.

Table 3.5   The target production time

| The target production time (Z) | | | | | | | | | | | | | | | |
|---|---|---|---|---|---|---|---|---|---|---|---|---|---|---|---|
| 1.5 | 1.6 | 1.7 | 1.8 | 1.9 | 2.0 | 2.1 | 2.2 | 2.3 | 2.4 | 2.5 | 2.6 | 2.7 | 2.8 | 2.9 | 3.0 |

Table 3.6   Relation between the number of processes and the optimal target production time

| Production period (n) | The optimal target production time | | |
|---|---|---|---|
| | Ascending order | Descending order | Optimal value |
| 2 | 2.2 | 2.6 | 2.2 |
| 3 | 2.0 | 2.3 | 2.2 |
| 4 | 1.7 | 2.1 | 1.7 |
| 5 | 1.5 | 1.9 | 1.5 |
| 6 | 1.4 | 1.7 | 1.3 |
| 7 | 1.2 | 1.5 | 1.2 |
| 8 | 1.2 | 1.6 | 1.1 |
| 9 | 1.1 | 1.3 | 1.0 |
| 10 | 1.0 | 1.3 | 0.9 |

In this case, the processing rates (the good processing rate is $\mu_s$ and the bad processing rate is $\mu_a$) of this particular worker are set as shown below.

$$\begin{aligned}
\text{(i)} \quad & \mu_s = 1.0, \mu_a = 0.1 \\
\text{(ii)} \quad & \mu_s = 1.5, \mu_a = 0.1 \\
\text{(iii)} \quad & \mu_s = 2.0, \mu_a = 0.1 \\
\text{(iv)} \quad & \mu_s = 1.1, \mu_a = 1.0
\end{aligned}$$

The other parameters are the same as above.

Table 3.7 shows the optimal assignments when one worker possessing a bad processing rate is part of the system. From Table 3.7, it can be noted that the expected cost of the system is minimum when the worker with a bad rate is arranged to be placed at the first place of the process.

Table 3.8 shows the optimal assignments when one worker possessing a good processing rate is a part of the system. From Table 3.8, it is found that optimal

**Table 3.7** The optimal assignment of a case where one worker possessing a bad processing rate is a part of the system

| Production period ($n$) | Optimal assignment |
|:---:|:---:|
| 2 | (A,S) |
| 3 | (A,S,S) |
| 4 | (A,S,S,S) |
| 5 | (A,S,S,S,S) |
| 6 | (A,S,S,S,S,S) |
| 7 | (A,S,S,S,S,S,S) |
| 8 | (A,S,S,S,S,S,S,S) |
| 9 | (A,S,S,S,S,S,S,S,S) |
| 10 | (A,S,S,S,S,S,S,S,S,S) |

**Table 3.8** The optimal assignment of a case in which one worker possessing a good processing rate is a part of the system

| Production period ($n$) | Optimal assignment |
|:---:|:---:|
| 2 | (A,S) |
| 3 | (A,S,A) |
| 4 | (A,A,S,A) |
| 5 | (A,A,S,A,A) |
| 6 | (A,A,A,S,A,A) |
| 7 | (A,A,A,S,A,A,A) |
| 8 | (A,A,A,A,S,A,A,A) |
| 9 | (A,A,A,A,S,A,A,A,A) |
| 10 | (A,A,A,A,A,S,A,A,A,A) |

assignments have a property such that the worker possessing a good rate is placed after the middle part of the process. The conclusions of Tables 3.7 and 3.8 can be obtained from the proposed algorithm in [14]. Finally, the recent development of this subject is available in Yamamot et al. [28].

# References

1. Conway, R. W., Maxwell, W. L., & Miller, L. W. (1967). *Theory of scheduling*. USA: Addison-Wesley.
2. Wild, R. (1972). *Mass-production management, the design and operation of production flow-line systems*. London: Wiley.
3. Matsui, M. (2011). Conveyor-like network and balancing, In A. B. Savarese (Ed.), Manufacturing engineering, Ch. 3. N.Y: NOVA.
4. Matsui, M. (2005). CSPS model: Look ahead controls and physics. *International Journal of Production Research, 43*(10), 2001–2025.
5. Matsui, M. (2014). *Manufacturing and service enterprise with risks II: The physics and economics of management*, Ch. 2. Tokyo: Springer.
6. Hopp, W. J., Spearman, M. L. (1995). *Factory physics*. McGraw-Hill.
7. Matsui, M. (2005). *Management of manufacturing enterprise: Profit maximization and factory science.* Kyuritsu-Shuppan (in Japanese).
8. Matsui, M. (2015). Development of factory science: Duality and balancing of job shop (FS) vs. convey or system(OE) by Matsui's flow approach. In *Proceedings of International Scheduling Symposium 2015, Kobe, Japan.*
9. Matsui, M. (2008). *Manufacturing and service enterprise with risks: A stochastic management approach.* N.Y.: Springer.
10. Matsui, M. (2005). A management cycle model: Switching control under lot processing and time span. *Journal of Japan Industrial Management Association, 56,* 256–264.
11. Yamamoto, H., Sun, J., & Matsui, M. (2010). A study on limited-cycle scheduling problem with multiple periods. *Computers & Industrial Engineering, 59,* 675–681.
12. Shore, B. (1995). *Operations management.* NY: McGraw-Hill.
13. Hitomi, K. (1979). *Manufacturing systems engineering.* London: Taylor & Francis.
14. Matsui, M. (2015). The invisible body-balancing economics: A medium approach. *Theoretical Economics Letters, 5*(1), 66–73.
15. Wight, Q. W. (1974). *Production and inventory management in the computer age.* Canners Publishing Co.
16. Verzijl, J. J. (1976). *Production planning and information systems* (pp. 16–17). London: Macmillan.
17. Benders, J. (2002). The origin of period batch control (PBC). *International Journal of Production Research, 40,* 1–6.
18. Bergamashi, D., Cigolin, R., Perona, M., & Portiol, A. (1997). Order review and release strategies in job shop environment: A review and a classification. *International Journal of Production Research, 35,* 399–420.
19. Swamidass, P. M. (2000). *Encyclopedia of production and manufacturing management.* Boston: Kliuwer.
20. Enns, S. T. (2001). MRP performance effects due to lot size and planned lead time settings. *International Journal of Production Research, 39,* 461–480.
21. Wu, X., & Zhou, X. (2008). Stochastic scheduling to minimize the expected maximum lateness. *European Journal of Operational Research, 190,* 103–115.
22. Matsui, M., Yamamoto, H., Liu, J. (2002). A theoretical study on limited-cycle problems. In *Reprints of Japan Industrial Management Association, Spring, B07* (in Japanese).

23. Yamamoto, H., Matsui, M., & Liu, J. (2006). A basic study on a limited-cycle problem with multi periods and the optimal assignment problem. *Journal of Japan Industrial Management Association, 57*, 23–31. (in Japanese).
24. Matsui, M., Shingu, T., Makabe, H. (1997). Conveyor-serviced production system: An analytic framework for station-centered approach by some queueing formulas. In *Preliminary Report of the Operations Research Society of Japan* (pp. 104–107) (in Japanese).
25. Hillier, F. S., & Boling, R. W. (1966). The effects of some design factors on the efficiency of production lines with variable operation times. *Journal of Industrial Engineering, 17*, 651–658.
26. Johnson, S. M. (1963). Optimal two-and three-stage production schedules with set-up times included. In J.F. Muth & G.L. Thompson (Eds.), *Industrial Scheduling*, Ch.2. Prentice-Hall.
27. Malon, D. M. (1984). Optimal consecuive-2-out-of-n: F component sequencing. *IEEE Transaction on Reliability*, R-33, 414–418.
28. Yamamoto, H., Kong, X., & Sun, J. (2015). Introduction to a limited-cycle with multiple periods. *Journal of Japan Industrial Management Association, 66*(2E), 169–181.

# Chapter 4
# Economics of Invisible Collaboration

**Abstract** Our global world consists of various arrangements in the division of work in nature and artifacts. We first present an invisible body-balancing/collaboration network and economics in an asymmetric society using a medium approach under cloud balancing. This medium approach to the economics of collaboration originates in the newsboy problem, and results from the invisible hand of the market (demand) speed with Chameleon's criteria. However, the invisible hand in economics causes a win–win situation (balancing) under a series type supply chain, but not necessarily for parallel types. It may be possible to address this balancing problem using supply chain management (SCM) ellipse theory, and to formulate an SCM/GDP network with the circulation of marginal profit (medium value) in economics. The ellipse theory of SCM refers to the win–win balance between economics and reliability. The former is attained by balancing even-costs, while this chapter shows that the latter holds for parallel chains by changing the focus from lead times to workloads. Finally, the economics of invisible collaboration is discussed using numerical examples and cases, and balancing principal is numerically verified based on Matsui's equation ($W = ZL$).

**Keywords** Collaboration/balancing · Medium approach · Ellipse theory · Invisible hand · SCM/GDP · Marginal profit/value

## 4.1 Medium/Invisible Collaboration of Bodies

### 4.1.1 Introduction to Collaboration

Global manufacturing involves many individuals participating in the division of work. This chapter considers the invisible body-balancing network and the associated economics using a medium approach. This medium approach to invisible collaboration originated in the Newsboy problem [1], and is dictated by the speed of demand (invisible hand).

© The Author(s) 2016
M. Matsui, *Fundamentals and Principles of Artifacts Science*,
SpringerBriefs in Business, DOI 10.1007/978-981-10-0473-5_4

The traditional balancing problem originates in the Ford system, and is essential to the economics of mass production in the automobile industry [2]. This problem is the speed of the demand-to-supply (cycle time) to the market, and relates to conveyor speed versus efficiency (cycle time) in the assembly line.

The related domain in Industrial Engineering (IE) is called line balancing and based on the principle of system balancing in the assembly industry, though this also includes service [3]. This solution trends to take the viewpoint of the mean inventory in the factory, the lean inventory on the line, and the speed in the supply chain (SCM) (demand-to-supply).

In the Toyota system [4], the line speed is determined and balanced by the speed of demand speed in the market, and the *Kanban* system solves the problem of efficiency versus *muda* (loss) in the pull-type demand and supply system. However, this solution aims to improve lean inventory.

We here consider the medium approach to the efficiency versus *muda* problem in stochastic system balancing [5], based on the medium inventory originating in the operations research (OR) newsboy problem. This section also applies this approach to the SCM/GDP system in a country-like region in the near future.

The section summarizes a recent potential approach to the medium (Chameleon's) balancing from conveyor systems to the SCM/GDP-economic system. First, it briefly outlines the body-balancing problem, and then presents a unified economic treatment of the physical balancing problem.

Finally, this section identifies and verifies an optimal balancing condition from the view of Matsui's equation and Chameleon's criteria. The results provide a shortcut to the traditional balancing method, for example, the 2-stage method [6]. This chapter uses the concept of balancing in physics and economics, and extends it to the science of the discrete world.

Further study shows that a progressive and autonomous control of marginal profit/value in GDP networks potentially influences a country's GDP. However, this requires a change from the invisible hand to the visible demand-to-supply approach.

## 4.1.2  Body-Balancing System and Medium Approach

### 4.1.2.1  Economics in the Balancing Problem

Modern society is a result of the worldwide division of work as we move toward globalization. In 1776, Smith [7] introduced the concept of the invisible hand, and this problem is more important in global economics. Classic economics put price at the center [7], and considered production quantity as the next actor [8]. Recently, Matsui [9] pointed out that demand speed (cycle time) is similar to the invisible hand, because it brings maximal-profit (re-)balancing in a changeable market economy.

Generally, the body-centered network (SCM/GDP) could be invisibly balanced and coordinated by the demand speed acting in a manner similar to the concept of

God's hand, plus cloud computing [10]. However, there is profit balancing for different types of series, and cost-relative balancing in parallel types otherwise [6]. Thus, this God-like hand would be alive and it is possible to attain win–win in not only SCM networks but also in Smith's world only under an equal partnership.

For example, consider the two- or three-center model consisting of sales, assembly, and fabrication centers [11]. The two main configuration types are series and parallel systems. The SCM is a series type, and enterprise resource planning (ERP) is a parallel type. For a series, profit maximization is attainable under the demand speed given or shared, even if each heterogeneous agent (enterprise) pursues its own goals without cooperation in the indivisible environment.

Thus, each unit-optimization in profit yields total optimization under noncooperation, and balancing occurs during the middle lead-time (reliability). This class is called the integral optimization, and might be governed by Matsui's Ellipse map and strategy [12].

Our profit corresponds to the marginal profit/value in accounting/GDP, and similar to the medium criterion [13]. In the classics, Smith presents the first concept of the invisible hand in 1759 [14], prior to 1776.

This concept might be similar to that of the medium criterion in our balancing problems, and would facilitate the invisible collaboration in Fig. 4.1. Figure 4.1a, b illustrate the basic and "*jyukyu*"(demand-to-supply) types of invisible collaboration, respectively.

### 4.1.2.2   Medium Balancing Approach

The stochastic balancing problem is a class of conveyor-serviced production station (CSPS) and its networks [15]. This is a station-centered approach to the power-conveyor system, and the station here corresponds to the individual bodies (human, house, enterprise, etc.) in a society or country.

The body-balancing system defines stochastic balancing as a stabling phenomenon of transient, bottlenecked versus balanced, state of the object system. The balanced solution with this stochastic balancing approach becomes quasi-optimal.

Figure 4.2 outlines a recent medium approach to SCM, the body-balancing system [13]. In Fig. 4.2, $d(>0)$ is the demand speed (cycle time), and $\beta_i$ and $MI_i$ are the medium criterion $(0 < \beta < 1)$, and moving-standard inventory, respectively, in the individual body $i(i = 1, 2, \ldots, n)$.

The medium criterion, $\beta$, is controlled by $\bar{\beta} = (\beta_3 - \beta_1)/(\beta_2 + \beta_3)$, in which the parameters $\beta_1, \beta_2$ and $\beta_3$ are the cost coefficients (penalties) of inventory holdings, excess inventory and shortage inventory, respectively.

Now, the Newsvendor's conditions [1] are the following:

$$F_i(MI_i) = \bar{\beta}_i, \quad i = 1, 2, \ldots, n \tag{4.1}$$

where $F_i(\cdot)$ is the distribution function of the inventory in the individual body, $i$.

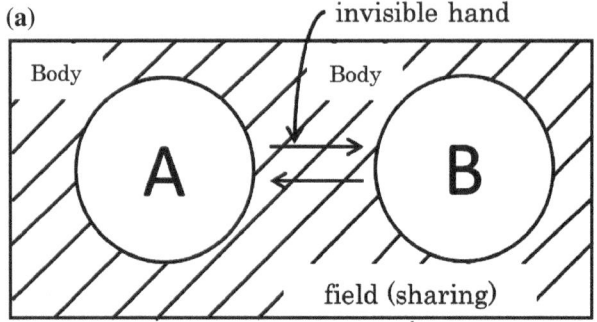

Basic invisible collaboration

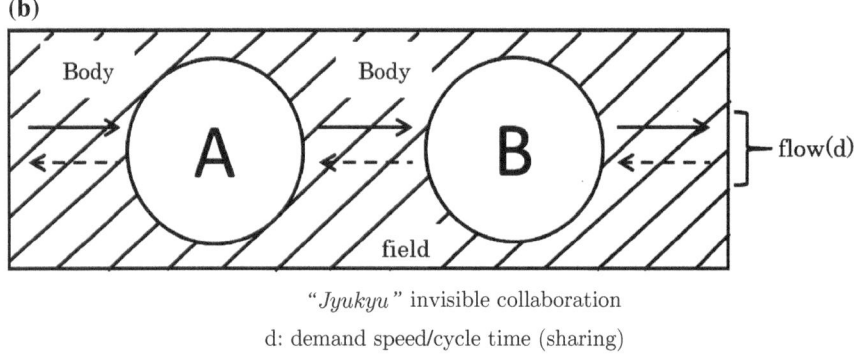

"*Jyukyu*" invisible collaboration

d: demand speed/cycle time (sharing)

$\longrightarrow$ : material,   $\leftarrow\!-\!-\!\cdot$ : information

**Fig. 4.1**  Model of invisible collaboration ($A$, $B$: heterogeneous bodies)

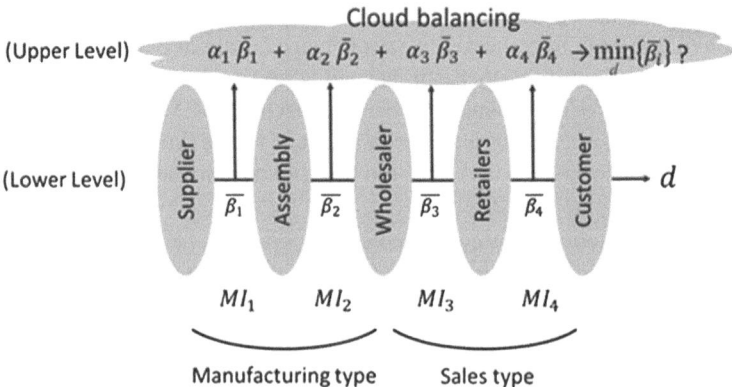

**Fig. 4.2**  A body-balancing system in a supply chain (economics) [9]

Then, the balancing goal is given by the objective function:

$$\alpha_1 \bar{\beta}_1 + \alpha_2 \bar{\beta}_2 + \cdots + \alpha_n \bar{\beta}_n \rightarrow \min_d \{\bar{\beta}_i\}, \tag{4.2}$$

where $a_i$ is the weight factor $(0 < a_i)$, $i = 1, 2, \ldots, n$.

An optimal condition (balancing) is assumed from the classical inequality and Matsui's equation $(W = ZL)$ [12] as follows:

$$\text{Hypothesis}: \alpha_1 \bar{\beta}_1 = \alpha_2 \bar{\beta}_2 = \cdots = \alpha_n \bar{\beta}_n = W(=ZL). \tag{4.3}$$

In (4.3), $Z$ and $L$ correspond to $a_i$ and $\bar{\beta}_i$, respectively, and $W$ represents a balancing value at the equilibrium.

## 4.1.3   Balancing Principles and Network

### 4.1.3.1   The d-Balancing Principle

Two principles of medium balancing are presented and considered here. First, the $d$-balancing problem (4.2) appears on the upper level of the 2-level scheme in Fig. 4.1. This main problem is easily decomposed into the dual problem:

$$F_i(I_i) = \bar{\beta}_i, \quad i = 1, 2, \ldots, n \tag{4.4}$$

in the body of entity $i$. Matsui's point, $\bar{\beta}_i$, is based on Chameleon's criteria [12].

Now, the following condition is considered under the demand speed (cycle time), $d$ $(0 < d < 1)$, and the exponential service with mean, $m_i$ (supply speed). That is,

$$G_i(d) = 1 - \exp\left(-d / m_i\right) = \bar{\beta}_i, \quad i = 1, 2, \ldots, n \tag{4.5}$$

and the demand speed, $d$, is

$$d = -m_i \ln\left(1 - \bar{\beta}_i\right). \tag{4.6}$$

On $d$-balancing, the following relation is obtained from (4.6):

$$m_i \ln\left(1 - \bar{\beta}_i\right) = m_j \ln\left(1 - \bar{\beta}_j\right), \quad i \neq j. \tag{4.7}$$

Especially, for Poisson service, the optimal condition is

$$F_i(I_i) = \sum_{i=1}^{I_i} P(d; m_i) = 1 - \bar{\beta}_i, \quad i = 1, 2, \ldots, n \tag{4.8}$$

where $P(\cdot)$ is a Poisson type distribution.

**Fig. 4.3** Outline of the
rebalancing problem and
Matsui's equation ($W = ZL$)

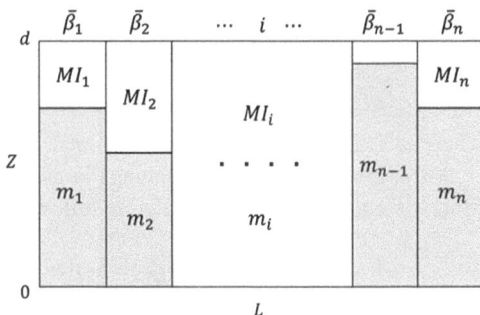

Figure 4.3 outlines these relationships. From Fig. 4.3 and Matsui's equation [13], the balancing equation is

$$\sum_{i=1}^{l_i} MI_i = nZ,\qquad(4.9)$$

and the balancing principle derives from (4.9) and the classic inequality is as follows:

$$L\sqrt{ZL} < \sum MI_i < ZL = W.\qquad(4.10)$$

#### 4.1.3.2  Network Flow Principle

On the lower level, the entity $(i, t)$ is the state of the body $i$ and period $t$, and Fig. 4.4 shows its network flow. This network flow would behave under demand speed, $d$, in body-centered balancing.

The following notations apply for each $i(=1, 2, \ldots, n)$ and $t(=0, 1, 2, \ldots)$ from [16]:

$O_{i,t}$  Order scheduled for body $i$ and period $t$,
$D_{i,t}$  Expected demand in body $i$ and period $t$,
$I_{i,t}$  Next inventory at the end of body $i$ and period $t$
   $(=I_{i,t}^+ - I_{i,t}^-)$,

where $I_{i,t}^-$ is the backorder position (quantity).

For each $t$, Fig. 4.4 and [16] provide the balancing relationship between demand and supply as follows:

$$O_{i,t} + I_{i-1,t}^+ + I_{i,t}^- = D_{i,t} + I_{i-1,t}^- + I_{i,t}^+, \quad i = 1, 2, \ldots, n\qquad(4.11)$$

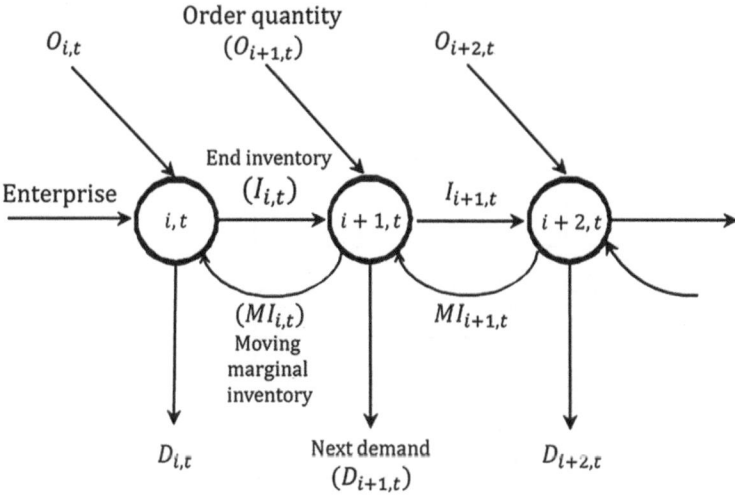

**Fig. 4.4**  SCM/GDP-economic system: a network flow with backorder

From (4.11), the second balancing principle is obtained:

$$O_{i,t} = D_{i,t} + \left(I_{i-1,t}^- - I_{i,t}^-\right) - \left(I_{i-1,t}^+ - I_{i,t}^+\right), \quad i = 1, 2, \ldots, n, \; t = 1, 2, \ldots \quad (4.12)$$

where the second and third terms on the right correspond to the coordination of the moving marginal inventory, $MI_{i,t}$, and end inventory, $I_{i,t}$, at $(i, t)$, respectively.

Equation (4.11) is similar to that for progressive control in [13]. Thus, our network flow is governed by $d$-balancing with order quantity, $O_{i,t}$, in (4.11).

## 4.1.4   d-Balancing

### 4.1.4.1   Conveyor System with Stopper

The usual conveyor systems are the two types of stations with or without a stopper. These systems are treated in [6] with a stochastic approach that helps minimize the irregular interruptions from delays and idleness in the series.

A cost approach to the delay and idleness in [6], as well as this balancing problem, is considered with the 2-stage method. Figure 4.5 illustrates a stopper type example, where the cycle time is 1.5 min, and the number of stations is $n = 5$.

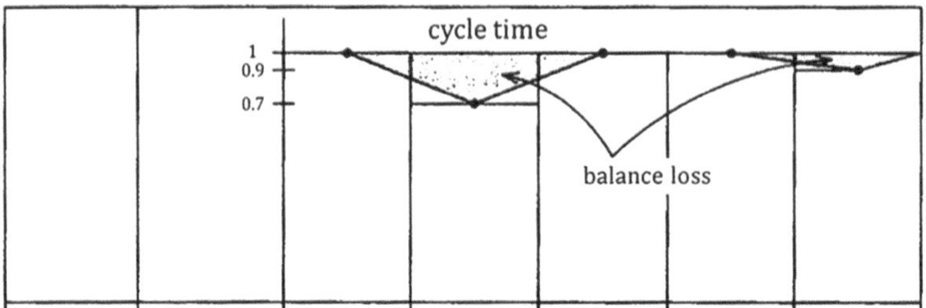

| | | cycle time | | | | |
|---|---|---|---|---|---|---|
| work arrangement | station | ① | ② | ③ | ④ | ⑤ |
| | assignment | (1,2,6) | (5,8) | (3,10) | (4,7) | (9,11) |
| | time | 1.0 | 0.7 | 1.0 | 1.0 | 0.9 | ( min ) |
| buffer | $N_i$ | 5 | 2 | 5 | 5 | 4 | $(d = 1.2)$ |
| | $C_i$ | 1.14 | 1.19 | 1.14 | 1.14 | 1.14 | $(d = 1.0)$ |

**Fig. 4.5** Pitch diagram and buffers: another verification

The newsboy method to this case [17] gives the penalty cost function as follows:

$$EC = \sum_{i=1}^{n} EC_i(d), \quad 0 < d < \infty \tag{4.13}$$

where the right side consists of the objective of $i$th station:

$$EC_i(d) = \beta_1 d + \beta_2 \int_0^d f_i(d)dx + \beta_3 \int_d^\infty f_i(d)dx, \tag{4.14}$$

in which $f(\cdot)$ is the p.d.f. of work time at the $i$th station.

Then, the optimal condition is obtained using the following differentiation method:

$$\sum_{i=1}^{n} f_i(d) = \frac{n\beta_1}{(\beta_3 - \beta_2)}. \tag{4.15}$$

The cycle time, $d$, is determined by Eq. (4.14), and this equation here replaces (4.2).

For example, let $\beta_1 = 5, \beta_2 = 10$ and $\beta_3 = 20$. From (4.15), the optimal cycle time, $d^*$, is given by $d^* = 1.4$ for the Erlangian with phase $k = 5$ in the station's work time.

Now, let us apply the SALPS (stochastic assembly line planner with strategy) soft simulator [13] to this case. Table 4.1 provides the simulation results with the optimal cycle time, $d'$. Thus, $d' = 1.3$ with a slight difference. In addition, the total cost is negligible.

**Table 4.1** Verification of medium balancing: simulation trial case

| $d$ | Total cost | $\bar{\beta}_1, \bar{\beta}_3, \bar{\beta}_4$ | $\bar{\beta}_2$ | $\bar{\beta}_5$ | $\sum \bar{\beta}_1$ |
|-----|-----------|--------------------------|--------|--------|------------|
| 1.0 | 469.5706 | 0.8773 | 0.6608 | 0.8541 | 4.146 |
| 1.1 | 91.20302 | 0.7790 | 0.6818 | 0.7973 | 3.816 |
| 1.2 | 85.43137 | 0.6692 | 0.6804 | 0.7200 | 3.408 |
| 1.3 | 84.78923 (SALPS) | 0.5591 | 0.6604 | 0.6324 | 2.970 |
| 1.4 | 84.88829 (optimal) | 0.4561 | 0.6260 | 0.5423 | 2.536 |
| 1.5 | 85.45898 | 0.3645 | 0.5813 | 0.4557 | 2.130 |
| 1.6 | 86.13179 | 0.2862 | 0.5300 | 0.3761 | 1.765 |

Figure 4.5 considers another verification, and shows a system bottleneck in the stations, 2 ($\bar{\beta}_2$) and 5 ($\bar{\beta}_5$), from Table 4.1. The balancing hypothesis in (4.3) is valid.

#### 4.1.4.2   Network Flow

Next, consider the franchise-type two-stage supply chain shown in Fig. 4.6. This case consists of the factory, $M$, and warehouse, $N$ [17].

In this case, the system is controlled with the inflow only, and the outflow is uncontrollable. Thus, the balancing problem is the minimization of total inventory between suppliers. That is,

$$\text{Net factory inventory}(M) + \text{Net warehouse inventory}(N) \rightarrow \min \{\bar{\beta}_i\}. \quad (4.16)$$

The objective Eq. (4.16) is here available to replace Eq. (4.2).

For this case, Table 4.2 presents and considers the $\bar{\beta}_1 - \bar{\beta}_2$ balancing table for network flow balancing. The total inventory in 1 year is 5,176, and a significant improvement compared to that in practice (8,074).

From Table 4.2, the medium balancing holds except for July, September, and April as follows:

$$\bar{\beta}_1, \bar{\beta}_2 = 0.4 - 0.5, \text{except July, September and April.} \quad (4.17)$$

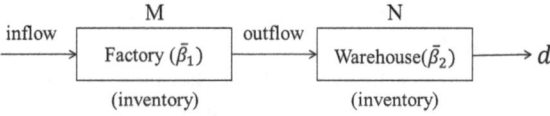

**Fig. 4.6**  A two-stage supply chain system to balance

**Table 4.2** $\bar{\beta}_1 - \bar{\beta}_2$ balancing table

| Month | 6 | 7 | 8 | 9 | 10 | 11 |
|---|---|---|---|---|---|---|
| Initial inventory | 7000 | 5595 | 7874 | 6632 | 5624 | 8782 |
| M-inventory $(\bar{\beta}_1)$ | 58(0.4) | 907(0.5) | 1010(0.6) | 2314(0.7) | 1418(0.5) | 897(0.5) |
| N-inventory $(\bar{\beta}_2)$ | 278(0.4) | 9(0) | 2057(0.5) | 1542(0.4) | 1607(0.4) | 866(0.4) |
| Outflow | 7723 | 7318 | 14,356 | 18,485 | 9122 | 15,245 |
| Inflow | 1059 | 2640 | 9549 | 15,710 | 6522 | 8227 |
| Inventory | 336 | 916 | 3067 | 3857 | 3025 | 1763 |
| Balance | 78,178 | 84,974 | 54,502 | 47,241 | 58,050 | 61,907 |
| Stock out | 0 | 0 | 0 | 0 | 0 | 0 |
| Month | 12 | 1 | 2 | 3 | 4 | 5 |
| Initial inventory | 6104 | 10,636 | 5856 | 13,343 | 12,626 | 7510 |
| M-inventory $(\bar{\beta}_1)$ | 1851(0.5) | 254(0.5) | 2577(0.5) | 91(0.3) | 2307(0.1) | 2837(0.5) |
| N-inventory $(\bar{\beta}_2)$ | 1490(0.4) | 1233(0.4) | 883(0.4) | 96(0.4) | 1386(0.4) | 2207(0.5) |
| Outflow | 11,201 | 19,609 | 13,748 | 23,422 | 13,055 | 9904 |
| Inflow | 8438 | 10,460 | 11,352 | 10,266 | 4122 | 6911 |
| Inventory | 3341 | 1487 | 3460 | 187 | 3693 | 4517 |
| Balance | 23,779 | 75,988 | 35,552 | 126,559 | 83,958 | 29,870 |
| Stock out | 0 | 0 | 0 | 0 | 0 | 0 |

Thus, it is doubtful that any bottleneck phenomena exists for July, September, and April. This result shows the effectiveness of the medium balancing method, and supports hypothesis (4.3).

## 4.2   Ellipse Map and SCM/GDP Collaboration

### 4.2.1   Ellipse Map and Collaboration

Our coordination approach deals with autonomous balancing by sharing demand (cycle time) information [18–22]. This chapter already examined a fundamental approach [23] from a station-centered perspective [24, 25], and a new challengeable trial already exists in a serial supply chain [18, 19, 22].

This section now extends the ellipse theory in supply chain research to a world of multiple win–win relationships. This problem also relates to an equilibrium/balancing condition and medium flow/value network with an invisible hand [26]. The main concern is with win–win balancing and how it creates maximum (marginal) GDP for all enterprises from an institutional point of view.

This invisible balancing in economics is again hypothesized to create win–win situations in the parallel type of SCM/GDP economics. Matsui [18] regarded this invisible hand as the demand speed, and the SCM ellipse theory considers the system balancing problem with demand speed.

A recent presentation described a medium approach to this invisible body-balancing economics approach [22]. The medium approach gives two efficiencies (lean "*Yase*") versus *muda* (fat "*Debu*") [19]. One is obtained by the newsvendor criteria in OR [22], and the other results from the marginal profit (value) in economics. This section uses and considers the latter.

Matsui [27] discussed a stochastic approach to system balancing or rebalancing using the latter medium demand speed based on parallel chain types, and presents an advanced theory to balance cost/profit with under and excess penalties based on demand speed. In conclusion, the ellipse theory of SCM referring to reliability and win–win balancing in economics also holds for parallel chains by changing the reliability from lead-time to workload.

## 4.2.2  Ellipse Theory and Medium Balancing

### 4.2.2.1  Fundamental Ellipse Theory of SCM

For the 2-center model, the ellipse theory for a job-shop was first proposed in 1983 [28] and developed to a class of management game model (MGM) between 2001 [29] and 2008 [18]. This fundamental economic theory is based on the following equation:

$$ER = EN + EC. \tag{4.18}$$

where $ER, EN$, and $EC$ refer to the mean revenue, (marginal) profit, and operating-cost (variable cost), respectively.

From Eq. (4.18), $ER^2 = EN^2 + EC^2$, and

$$\frac{EC^2}{a} + \frac{EN^2}{b} = 1, \tag{4.19}$$

where $p$ and $q$ are parameters, and Eq. (4.19) shows the standard type of ellipse in Fig. 4.7a.

Additionally, for the ellipse showing reliability, the fundamental theory is based on the following equation:

$$W = ZL = DL + mL, \quad Z(d) = D + m. \tag{4.20}$$

where $d$ is demand speed, $m$ is processing time, and $D$ is delay time.

From (4.20), $W^2 = Z^2 L^2 = (DL)^2 + (mL)^2$, and

$$\frac{(DL)^2}{p} + \frac{(mL)^2}{q} = 1, \tag{4.21}$$

where $p$ and $q$ are parameters, and Eq. (4.21) shows the standard type of ellipse in Fig. 4.7b.

**Fig. 4.7** Fundamentals of ellipse theory: economics versus reliability

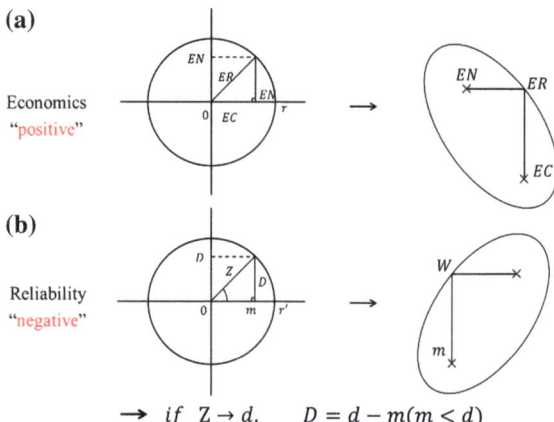

Furthermore, for the 2-chain scenario, the composite of two circles (enterprises) becomes elliptical and are adopted here as the ellipse theory of SCM.

### 4.2.2.2  Medium Balancing and Hypothesis

The ideal goal of entries (enterprises, homes, etc.) is integral system balancing, meaning that both economics (profit) and reliability (lead-time) hold in a win–win balance. In economics, the objective criterion is the positive sum of each (marginal) profit in the enterprise.

For a 2-center scenario, the integral profit in the medium approach is:

$$DEN = (ER_1 - EC_1)^+ + (ER_2 - EC_2)^+ \rightarrow \max, \atop c \text{ or } \lambda \tag{4.22}$$

where $ER_i$ and $EC_i$ are the revenue and operating-cost per unit time of enterprise $i$ (=1, 2), respectively. The difference $(ER_i - EC_i)$ corresponds to the marginal profit, $EN_i$, in the efficiency versus *muda* criteria of bodies.

For reliability, the mean workload is used as another balancing measure to replace lead time, and is given by

$$BT_i = L_i/\mu_i, \quad i = 1, 2 \tag{4.23}$$

where $L_i$ is the mean number of units, and $\mu_i$ is the processing rate. The objective criterion for a 2-center scenario is an optimization of the sum:

$$BT = BT_1 + BT_2 \rightarrow \min. \atop c \text{ or } \lambda \tag{4.24}$$

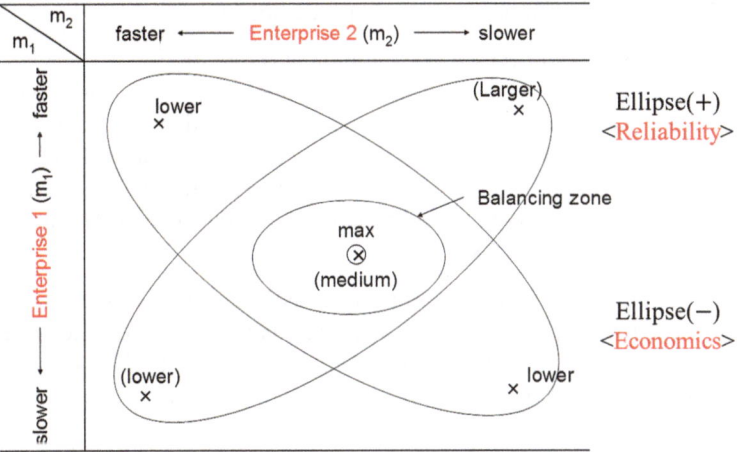

**Fig. 4.8** SCM ellipse hypothesis

The two economic criteria (4.22) and reliability (4.23) yield the integral optimization in another type of medium balancing (Fig. 4.8). This evidence is shown later in the ellipse theory of parallel chains.

Moreover, the dual type of *DEN* (4.22), *MEN*, is as follows:

$$MEN = (ER_1 + ER_2) - (EC_1 + EC_2) \tag{4.25}$$

Similar to reliability theory, the win–win balancing satisfies Eq. (4.22) with the following condition:

$$if\ EC_1 = EC_2, \quad EN_1 = EN_2. \tag{4.26}$$

### 4.2.3 Parallel SCM Balancing

#### 4.2.3.1 In 2-Chain Manufacturing

For this study, we present two parallel models consisting of heterogeneous enterprises (agents) [18, 30], assuming a different institutional environment from those in past bodies. One parallel model is make-or-buy manufacturing, and the other is a sales type with supplier–retailers.

The first model consists of two make-to-order (MTO) enterprises as follows: suppose that Job-shop 1 is a high cost domestic shop, while Job-shop 2 is a low cost shop in China. Profitable orders are accepted at Job-shop 1 and rejected orders are accepted at Job-shop 2.

The following assumptions and notations are added:

1. The orders arrive with Poisson distribution at a rate of $\lambda$.
2. The orders' marginal profit, $S$, has an exponential distribution with a mean of 1.
3. The shops' processing time is an exponential distribution with rates $\mu_1$ and $\mu_2$.
4. At Job-shop 1, the arriving orders are screened until the stock level reaches a backlog, $N$, by selection criterion, $c$ ($0 \le c \le \infty$). The rejected orders are sent to Job-shop 2 and accepted until reaching a backlog stock level of $M$. The overflow rate from Enterprise 2, $v$, is lost.

Thus, Job-shop 1 makes the make-or-buy decision and acts without a return from selection criterion (input speed), $c$, and may have a backlog stock level of $N$. Job-shop 2 may have a backlog stock level, $M$, and if the backlogs is grater than $M$, then a new order arriving after this point is lost. Job-shop 2 communicates with Job-shop 1, but both are in a noncooperative relationship.

Using the steady-state probabilities $\pi(n, m)'$ the objective functions for production are easily obtained [18, 31], and the mean backlogs, $J_1$ and $J_2$, are

$$J_1 = \sum_{m=1}^{M} \sum_{n=0}^{N} n\pi(n, m), \tag{4.27}$$

$$J_2 = \sum_{n=0}^{N} \sum_{m=1}^{M} m\pi(n, m), \tag{4.28}$$

in Job-shops 1 and 2, respectively.

In addition, the mean workloads of job-shops, $BT_1$ and $BT_2$, are given directly by Eq. (4.23). These objective criteria are used as the balancing measure for the two job-shops.

### 4.2.3.2  2-Chain Retailer Type

The second model consists of a supplier and two retailers with the following assumptions and notations:

1. The retailers' demand patterns have Poisson distributions with rates $\mu_1$ and $\mu_2$, respectively
2. The retailers sell the goods at price $p_2$. If each retailers' stock is sold out, the customer is lost to both retailers. The number of lost items is denoted by $K_i, i = 1, 2$.
3. The truck is first routed from the supplier to Retailer 1, and replenishes to stock level, $N$, before moving to Retailer 2 with a negligible delay.
4. The overflow rate from Retailer 2, $v$, is returned to the supplier.

The supplier has an infinite capacity, but the two retailers have the stock levels $N$ and $M$ as vendor managed inventory (VMI). Additionally, the supplier has a truck with travel time $R$, and replenishes the retailers' goods (at price $p_1$) at the approximate rate (input speed), $\lambda$.

Similar to the case in Sect. 4.2.3.1, the objective functions for the sales type are easily obtained using the steady-state probabilities $\pi(n, m)'$ in [18, 31], and the retailers' mean workloads, $BT_1$ and $BT_2$, are given instead of lead time by (4.23). These objective criteria are used as the balancing measure for the two retailers.

## 4.2.4 Balancing of Parallel Types

### 4.2.4.1 Make-to-Order and Retailers

Figure 4.9 reports the MTO balancing table, with ranges $0.2 < c < 3.0$ and $1 < N < 6$. From Fig. 4.9, the balance table shows the SCM ellipse theory in both economics and reliability (workload). In addition, the integral optimization shows both the even-cost and even-workload.

Similar to Fig. 4.9, Fig. 4.10 shows the retailers' balance table with ranges $1.0 < \lambda < 4.0$ and $1 < N < 6$. From Fig. 4.10, the balance table shows the SCM ellipse theory in both economics and reliability (workload). It is here noted that the workload balance is still unsolved.

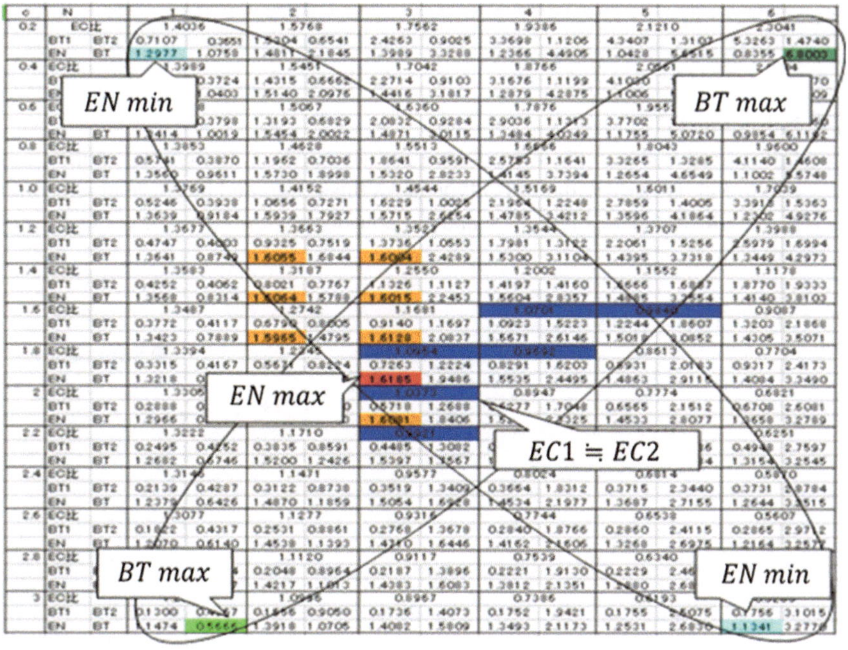

**Fig. 4.9** Balance table: MTO type ($\mu_1 = 6$, $\mu_2 = 6$)

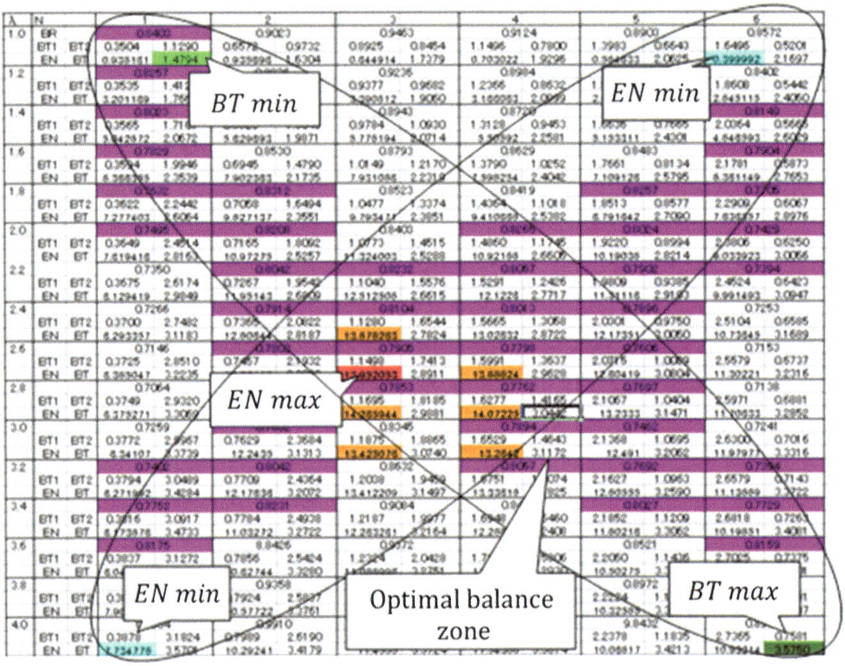

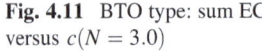
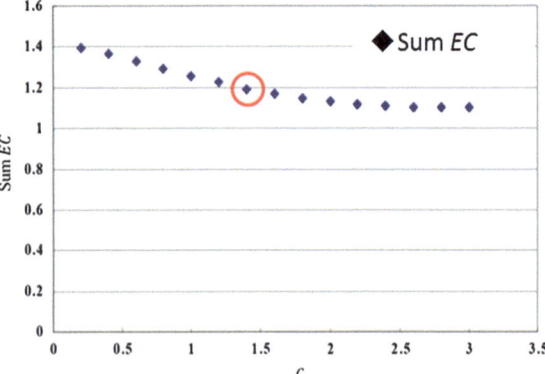

**Fig. 4.10** Balance table: retailer type ($\mu_1 = 6, \mu_2 = 6$)

#### 4.2.4.2  Parallel Balancing Issues

(a)  Balancing MTO type

In the earlier study, the total chain cost (Sum $EC$) was minimal at the profit maximum. However, the MTO type lacks this point, and has instead a medium point, illustrated in Figs. 4.11 and 4.12 at downward curve in c and upward curve in $N$.

**Fig. 4.11** BTO type: sum EC versus $c (N = 3.0)$

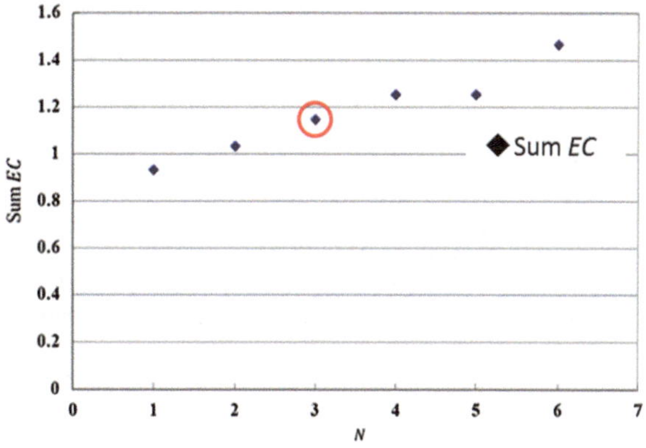


**Fig. 4.12** BTO type: sum EC versus $N(c = 1.8)$

(b) Retailer balancing
For heterogeneous retailers, the integral or balancing optimization is incomplete. Although the outcomes related to the ellipse theory are visible for the economic factor, Fig. 4.13 shows that this does not exist for the other factor for heterogeneous retailers.

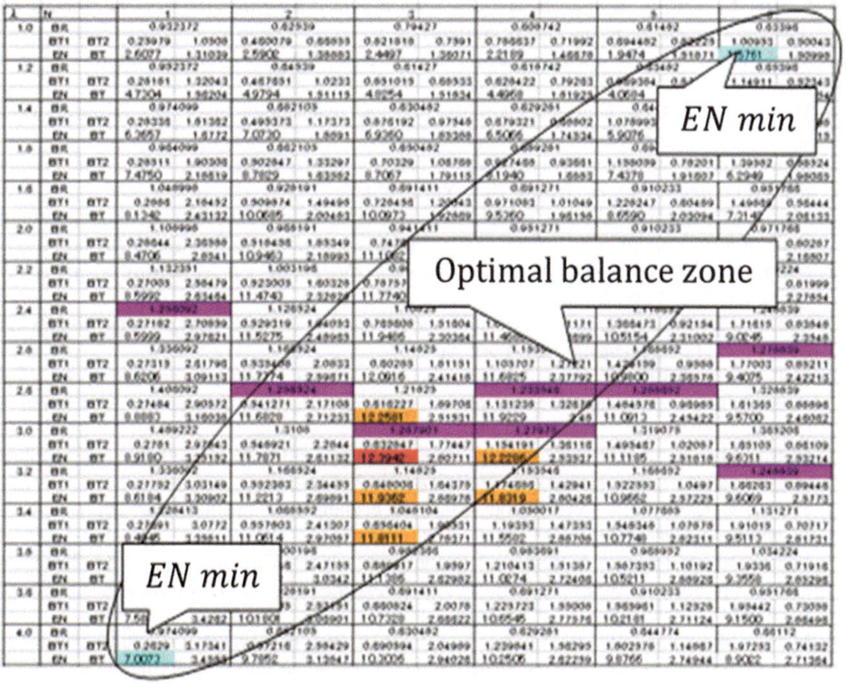


**Fig. 4.13** Heterogeneous retailers' balance table ($\mu_1 = 8, \mu_2 = 6$)

# References

1. Weeks, J. K. (1979). Optimizing planned lead time and delivery dates. In *21st Annual Conference Proceedings*, APICS, pp. 177–188.
2. Wild. R. (1972). *Mass-production management, the design and operation of production flow-line systems*. London: Wiley.
3. Stamp, D. (1995). *The invisible assembly line-boosting white-collar productivity in the new economy*. New York: AMACCM.
4. Ohno, T. (1988). *Toyota production system: Beyond large-scale*. Productivity Press.
5. Matsui, M. (2015). The invisible body-balancing economics: A medium approach. *Theoretical Economics Letters, 5*, 66–73.
6. Matsui, M. (2008). *Manufacturing and service enterprise with risks: A stochastic management approach*. New York: Springer.
7. Smith, A. (1776). *The wealth of nations* (1952 ed.). Encyclopedia Britannica.
8. Keynes, J. M. (1936). *The general theory of employment, interest and money* (1973 ed.). Macmillan Press.
9. Matsui, M. (2010). Division of work, stochastic (re-)balancing and demand speed: From assembly line toward demand chain. *Journal of Japan Industrial Management Association (JIMA), 60*(6E), 324–330.
10. Matsui, M. (2013). Modern God-like hand: Cloud balancing issues on global SCM networks and economics. In *Euro-Asia Economic Forum (EAEF 2013)*, Xi'an, China, p. 285.
11. Matsui, M. (2012). Economic station-centered network and invisible collaboration: A cyclic vs. semi-cyclic view. *Theoretical Economics Letters, 2*, 344–349.
12. Matsui, M. (2012). Economic demand-balancing problem of multi-center. In A. Tavdize (Ed.), *Progress in economics research* (Vol. 25, Ch. 12, pp. 227–235). New York: NOVA.
13. Matsui, M. (2014). *Manufacturing and service enterprise with risks II: The physics and economics of management*. Tokyo: Springer.
14. Smith, A. (1959). *The theory of moral sentiments* (1970 ed.). London: Oxford University Press.
15. Matsui, M. (2011). Conveyor-like network and balancing. In A. B. Savarese (Ed.), *Manufacturing engineering* (Ch. 3, pp. 65–87). New York: NOVA.
16. Johnson, L. A., & Montgomery, D. C. (1974). *Operations research in production planning, scheduling and inventory control*. New York: Wiley.
17. Matsui, M. (2014). Medium (Chameleon's) balancing: From conveyor system toward supply chain. *Reprints of Japan Industrial Management Association*, Spring, 16–21 (in Japanese).
18. Matsui, M. (2008). *Manufacturing and service enterprise with risks: A stochastic management approach*. New York: Springer.
19. Matsui, M. (2014). *Manufacturing and service enterprise with risks II: The physics and economics of management*. New York: Springer.
20. Matsui, M. (2010). Division of work, stochastic (re-)balancing and demand speed: From assembly line toward demand chain. *Journal of Japan Industrial Management Association, 60*(6E), 324–330.
21. Matsui, M. (2012). Economic station-centered network and invisible collaboration: cyclic vs semi-cyclic view. *Theoretical Economics Letters, 2*, 344–349.
22. Matsui, M. (2015). The invisible body-balancing economics: A medium approach. *Theoretical Economics Letters, 5*(1), 66–73.
23. Matsui, M., & Ichihara, S. (2003). A game approach to SCM: Modeling, formulation and integral optimization. In *Proceedings of 17th International Conference Production Research*, Virginia, U.S.A.
24. Matsui, M., Shingu, T., & Makabe, H. (1977). Conveyor-serviced production system: An analytic framework for station-centered approach by some queueing formulas. *Preliminary Report of the Operations Research Society of Japan*, Autumn, 104–107 (in Japanese).

25. Matsui, M. (1982). Conveyor-serviced production system (II): 2-level mathematical formulation and application. *Preliminary Report of the Operations Research Society of Japan*, Autumn, 92–93 (in Japanese).
26. Smith, A. (1776). *The wealth of nations* (1952 ed., p. 194). Encyclopedia Britannica.
27. Matsui, M. (2015). Economics of collaboration: Another medium flow/value approach to the invisible body-balancing economics for SCM/GDP. In *Proceedings of 23rd International Conference on Production Research*, Manila, Philippines.
28. Matsui, M. (1983). A game-theoretic consideration of order-selection and switch-over policy. In *Preprints of Japan Industrial Management Association*, Fall Meeting, pp. 48–49 (in Japanese).
29. Matsui, M. (2001). A management game model: Economic traffic, leadtime and pricing setting. *Journal of Japan Industrial Management Association, 53*, 1–9.
30. Iizuka, H., Matsui, M. (2001). A profit-balancing consideration for K-center model of series-parallel type (2). *Preliminary report of Japan Operations Research Society* (pp. 204–205). Spring (in Japanese).
31. Matsui, M., Dai, Y. (2006). Parallel SCM: Balancing of make-or-buy and retailer types. In *International Workshop on Institutional View of Supply Chain Management*. Japan: Tokyo Institute of Technology.

# Appendix

## A.1 Newtonian Mechanics (Point Mass) Versus Matsui's Dynamics (Body)

Generally, Newtonian mechanics is the basic principle applied to a point mass, while Matsui's dynamics is the basic principle in 3M&I-body. Both nature and artifacts correspond in three points and can be unified as follows:

(1) First law (law of inertia)

$$\text{Lead time} : W = ZL \,(\text{stationary})$$

$$\leftrightarrow \text{Momentum} : p = mv (\text{inertia}),$$

(2) Second law (equation of motion)

$$\text{Fund} : \bar{W}(= Z\bar{L} = ZW) = Z^2 L$$

$$\leftrightarrow \text{Motion} : F = mv^2,$$

(3) Third law (action and reaction)

$$\text{Action}\,(\vec{F}_{12}, \vec{W}_{12})$$

$$\leftrightarrow \text{Reaction}\,(\vec{F}_{21}, \vec{W}_{21}) : \text{Kepler's ellipse theory},$$

where both notations are similar as below:

$$\text{speed}(v) \leftrightarrow \text{unit cycle}(Z) : \text{integral},$$

$$\text{mass}(m) \leftrightarrow \text{inventory}(L) : \text{position potential}.$$

© The Author(s) 2016
M. Matsui, *Fundamentals and Principles of Artifacts Science*,
SpringerBriefs in Business, DOI 10.1007/978-981-10-0473-5

## A.2  Historical Review of 3M&I-body Science (Since the Time of Archimedes)

The historical view of 3M&I-body science is outlined in the following figure. Now, the two principles of balancing and sandwich are proposed and developed in this book toward nature versus artifacts theory.

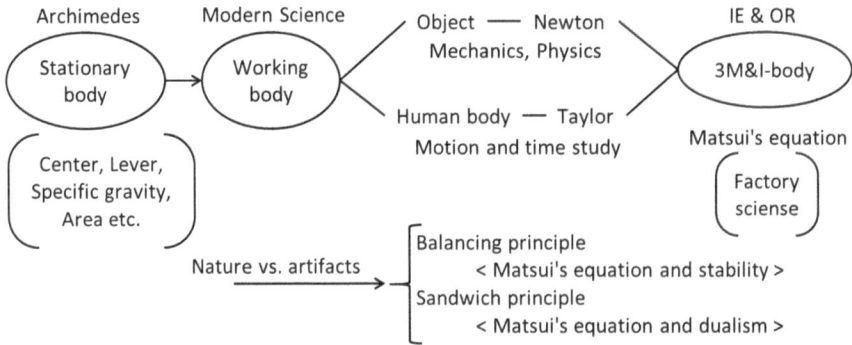

*Remark* The father of management, P.F. Drucker, points out as follows:

"Taylor, though the Isaac Newton (or perhaps the Archimedes) of the science of work, laid only first foundations, however. Not much has been added to them since—even though he has been dead all of sixty years." in "Management: Tasks, Responsibilities, Practices" (Harper and Row 1974)

# Index

© The Author(s) 2016
M. Matsui, *Fundamentals and Principles of Artifacts Science*,
SpringerBriefs in Business, DOI 10.1007/978-981-10-0473-5